Praise for Gwynne Dyer and *Climate Wars*

NATIONAL BESTSELLER

"Dyer is an accomplished military historian who bolsters his extensive knowledge with a rhetorical style that is at once invisible and entirely convincing."
Publishers Weekly

"Meticulously collected and expertly analyzed. And while the potential outcomes may seem dire, there are points of hope presented in the book. Hopefully, people are reading."
Popjournalism

"One of Canada's best known defence and foreign policy analysts."
National Post

"Dyer's style of supporting glib editorial comment with incessant statistics, creates a chilling effect of the near-inevitability of climate wars."
Courier-Island

GWYNNE DYER

Also by Gwynne Dyer

War
The Mess they Made: The Middle East After Iraq
With Every Mistake
Future Tense: The Coming World Order
Ignorant Armies: Sliding into War in Iraq

CLIMATE WARS

VINTAGE CANADA

VINTAGE CANADA EDITION, 2009

Published in Canada by Vintage Canada, a division of Random House of Canada Limited, in 2009. Originally published in hardcover in Canada by Random House Canada, a division of Random House of Canada Limited, in 2008. Distributed in Canada by Random House of Canada Limited, Toronto.

Vintage Canada and colophon are registered trademarks of Random House of Canada Limited.

www.randomhouse.ca

LIBRARY AND ARCHIVES CANADA CATALOGUING IN PUBLICATION

Dyer, Gwynne
    Climate wars / Gwynne Dyer.

Includes index.
ISBN 978-0-307-35584-3

    1. Global warming—Political aspects. 2. Climatic changes—Political aspects. 3. Global warming—Social aspects. 4. Climatic changes—Social aspects. 5. Climatic changes—Regional disparities. I. Title.

QC981.8.C5D94 2009   363.738'74   C2009-900364-3

Book design by Andrew Roberts

Printed and bound in the United States of America

10 9 8 7 6 5 4 3 2 1

For James, who will spend his whole life dealing with this. And of course, for Caro and Flor.

Contents

Introduction

Recent scientific evidence has . . . given us a picture of
the physical impacts on our world that we can expect as
our climate changes. And those impacts go far beyond
the environmental. Their consequences reach to the
very heart of the security agenda.

—Margaret Beckett,
former British foreign secretary

THIS BOOK IS AN ATTEMPT, peering through a glass darkly,
to understand the politics and the strategies of the potentially
apocalyptic crisis that looks set to occupy most of the twenty-
first century. There are now many books available that deal
with the science of climate change and some that suggest possible approaches to getting the problem under control, but
there are few that venture very far into the grim detail of how
real countries experiencing very different and, in some cases,
overwhelming pressures as global warming proceeds, are
likely to respond to the changes. Yet we all know that it's
mostly politics, national and international, that will decide
the outcomes.

Two things in particular persuaded me that it was time
to write this book. One was the realization that the first and
most important impact of climate change on human civilization will be an acute and permanent crisis of food supply.
Eating regularly is a non-negotiable activity, and countries

that cannot feed their people are unlikely to be "reasonable" about it. Not all of them will be in what we used to call the "Third World"—the developing countries of Asia, Africa and Latin America.

The other thing that finally got the donkey's attention was a dawning awareness that, in a number of the great powers, climate change scenarios are already playing a large and increasing role in the military planning process. Rationally, you would expect this to be the case, because each country pays its professional military establishment to identify and counter "threats" to its security, but the implications of their scenarios are still alarming. There is a probability of wars, including even nuclear wars, if temperatures rise two to three degrees Celsius. Once that happens, all hope of international cooperation to curb emissions and stop the warming goes out the window.

The activists who warn about the consequences of climate change generally do not go too deeply into these issues in public, perhaps for fear of inducing despair in those whom they seek to mobilize. Back in the days when the climate-change denial industry was still manufacturing a fake "debate" to cast doubt on the whole phenomenon of global warming, especially in North America, there may have been a tactical case for soft-pedalling the consequences for fear of sounding too extreme. (Although I don't think so, really. It almost always turns out to be better in the end to state the facts as you see them, and let the chips fall where they may.) At any rate, the denial industry is now in full retreat, and it's high time for everybody to say exactly what they mean.

As this is a book about the political and strategic consequences of climate change, for the basic science and some of the commoner physical global-warming scenarios I have depended mainly on published secondary sources, such as the 2007 *Intergovernmental Panel on Climate Change Fourth Assessment Report* and the 2006 *Stern Review on the Economics of Climate Change,* supplemented and updated with interviews where necessary. Indeed, in some ways the interviews are the real foundation of the book: trekking around a dozen countries talking to the scientists, soldiers, bureaucrats and politicians who are immersed in these issues on a daily basis has been an enlightening experience, and one that did much to restore my trust in the rationality of human beings.

I have not bothered to revisit the arguments that once had to be made to persuade people that human civilization is having a serious effect on the climate, on the grounds that this debate has now ended, except among a few diehard sceptics. There must remain some infinitesimal possibility that the sceptics are right and everybody else is wrong, but the evidence for global warming caused by human activities is so strong that urgent action is required. The potential cost of doing too little, too late is vastly greater than the cost that might be incurred by doing more to fight global warming than turns out, at some later date, to have been strictly necessary.

The scenarios that precede each chapter are not intended to be predictions, but only examples of the kinds of political crisis that could be caused by climate change. Neither are they components in some larger vision of how the future will unfold; each stands alone, and it is of no importance if one should contradict another. When I quote experts from interviews or other sources in these scenarios, it is solely to illustrate that some assumption I am making is regarded as plausible by the experts, and does not imply that the person quoted agrees with or has even seen the scenario in question. The dates I have assigned to the various scenarios are particularly arbitrary, and could easily be pushed down several decades if global warming proceeds more slowly than the latest evidence seems to suggest. All that said, I have tried to make the scenarios as credible as possible, drawing on a lifetime of analyzing how the world works as an international-affairs journalist. Sometimes I even got it right.

And here, right up front, are four conclusions that I have reached after a year of trailing around the world of climate change—four important things that I did not fully understand when I started this trip. First, this thing is coming at us a whole lot faster than the publicly acknowledged wisdom has it. When you talk to the people at the sharp end of the climate business, scientists and policy-makers alike, there is an air of suppressed panic in many of the conversations. We are not going to get through this without taking a lot of casualties, if we get through it at all.

Second, all the stuff about changing the light-bulbs and driving less, although it is useful for raising consciousness and gives people some sense of control over their fate, is practically irrelevant to the outcome of this crisis. We have to decarbonise our economies wholesale, and if we haven't reached zero greenhouse gas emissions globally by 2050—and, preferably, 80 percent cuts by 2030—then the second half of this century will not be a time you would choose to live in. If we have done it right, on the other hand, then the fuel that runs our cars and planes, like the power that lights our homes and drives our industries, will not produce carbon dioxide or other greenhouse gases. Use as much as you want, or can afford.

Third, it is unrealistic to believe that we are really going to make those deadlines. Maybe if we had got serious about climate change fifteen years ago, or even ten, we might have had a chance, but it's too late now. Global greenhouse gas emissions were rising at about one percent a year when the original climate change treaty was signed in 1992; now they are growing at three percent a year, and most of Asia, home to half of the human race, is rapidly moving into industrialized consumer societies. To keep the global average temperature low enough to avoid hitting some really ugly feedbacks, we need greenhouse gas emissions to be falling by four percent NOW, and you just can't turn the supertanker around that fast. So we are going to need geo-engineering solutions as stopgaps to hold the temperature down while we work at getting our emissions down, and we should be urgently examining our options in this area now. There is a very broad consensus that we should not even discuss geo-engineering techniques because of the "moral hazard" they represent—because we might choose geo-engineering methods INSTEAD OF emissions reductions—but we get only one shot at solving this problem, and we will probably fail without geo-engineering.

And fourth, for every degree that the average global temperature rises, so do the mass movements of population, the number of failed and failing states, and very probably the internal and international wars. Which, if they become big and frequent enough, will sabotage the global cooperation that is the only way to stop the temperature from continuing to climb.

I should mention, finally, that I am known in some circles for having worried aloud at some length about the threat of nuclear war during the bad old days of the Cold War. (So much so that one well-known environmentalist, whom I have known for half my life, accused me of "hopping aboard" the issue of climate change when I asked him for an interview, as if I hadn't paid for the tickets to ride his train.) But the threat of nuclear war and nuclear winter that hung over the late twentieth century, the danger of runaway climate change that besets us now, and the unknown but predictably terrifying crises that will imperil our children and our grandchildren even if we stop global warming, are all facets of the same basic truth: as a species, we have achieved critical mass.

There are now so many of us and we consume so much that it would take about thirty percent more planet than we actually have to support us sustainably. If all the rest of the world's people attained a "Western" standard of living, we would need three to four planets. This particular crisis about the climate is soluble, mainly by moving on from the technologies of the Industrial Revolution to ones that are less crude and less damaging environmentally, but our powers have grown so great that in a larger sense the crisis is now permanent, although its specific character will change from time to time. We may not even have the luxury of having to confront only one apocalyptic crisis at a time (although that would be nice). For example, by the 2020s, we may be plunged into a struggle over the proper role of artificial intelligence that is just as important to the future of the human race as getting our impact on the climate under control. And out beyond the "known unknowns," as former U.S. secretary of defense Don Rumsfeld put it, lie the "unknown unknowns."

As the petty officer who dominated my life during navy basic training used to say: "If you can't take a joke, you shouldn't have joined."

THE YEAR 2045

Average global temperature: 2.8 degrees Celsius higher than 1990.

Global population: 5.8 billion.

SINCE THE FINAL COLLAPSE OF THE EUROPEAN UNION in 2036, under the stress of mass migration from the southern to the northern members, the reconfigured Northern Union (France, Benelux, Germany, Scandinavia, Poland and the old Habsburg domains) has succeeded in closing its borders to any further refugees from the famine-stricken Mediterranean countries. Italy, south of Rome, has been largely overrun by refugees from even harder-hit North African countries and is no longer part of an organized state, but Spain, Padania (northern Italy) and Turkey have all acquired nuclear weapons and are seeking (with little success) to enforce food sharing on the better-fed countries of northern Europe. Britain, which has managed to make itself just about self-sufficient in food by dint of a great national effort, has withdrawn from the continent and shelters behind its enhanced nuclear deterrent.

Russia, the greatest beneficiary of climate change in terms of food production, is the undisputed great power of Asia. However, the reunification of China after the chaos of the 2020s and 2030s poses a renewed threat to its Siberian borders, for even the much reduced Chinese population of eight hundred million is unable to feed itself from the country's increasingly arid farmland, which was devastated by the decline of rainfall over the north Chinese plain and the collapse of the major river systems. Southern India is re-emerging as a major regional power, but what used to be northern India, Pakistan and Bangladesh remain swept by famine and anarchy, due to the collapse of the flow in the glacier-fed Indus, Ganges and Brahmaputra rivers and the increasingly frequent failure of the monsoon. Japan, like Britain, has withdrawn from its continent and is an island of relative prosperity bristling with nuclear weapons.

The population of the Islamic Republic of Arabia, which had risen to forty million, fell by half in five years after the exhaustion of the giant Ghawar oil field in 2020, and has since halved again due to the exorbitant price of what little food remains available for import from any source. Uganda's population, 5 million at independence in 1962, reached 110 million in 2030 before falling back to 30 million, and the majority are severely malnourished. Brazil and Argentina still manage to feed themselves, but Mexico has been expelled from the North American Free Trade Area, leaving the United States and Canada with just enough food and water to maintain at least a shadow of their former lifestyles. The Wall along the U.S.-Mexican border is still holding.

Human greenhouse-gas emissions temporarily peaked in 2032, at 47 percent higher than 1990, due largely to the dwindling oil supply and the Chinese Civil War. However, the release of thousands of megatons of methane and carbon dioxide from the melting permafrost in Arctic Canada, Alaska and Siberia has totally overwhelmed human emissions cuts, and the process has slid beyond human ability to control. The combined total of human and "neo-natural" greenhouse gas emissions continues to rise rapidly, and the average global temperature at the end of the century is predicted to be eight or nine degrees Celsius higher than 1990.

Prognosis: Awful.

CHAPTER ONE
The Geopolitics of Climate Change

THE SCENARIO I'VE JUST DESCRIBED is not the sort that the climate modellers produce; they wisely stay well clear of any attempt to describe the political, demographic and strategic impacts of the changes they foresee. My scenario also posits a higher global average temperature for 2045 than the bulk of the models predict, but 2.8 degrees Celsius higher by that date is within the range of possibility, especially if some of the positive feedback mechanisms, such as the partial failure of the oceanic carbon sinks, the melting of the permafrost, and an ice-free Arctic Ocean in the summertime, begin to operate within this period. Unhappily, recent data from the tropical oceans, the permafrost belt and the Arctic Ocean suggest that all these feedbacks may be starting to kick in now, much earlier than expected.

The scenario also assumes that the governments of the planet will not have taken advantage of the twenty-year window of opportunity that we still have to get global emissions of greenhouse gases down by 80 percent. It assumes that mid-century will see the world on the upper path of global heating, with the planet's average temperature already two or three degrees Celsius hotter and heading for eight, nine or ten degrees hotter by century's end. In this world, our worries are not just hotter summers, bigger hurricanes, rising sea levels and polar bears swimming for their lives. We are trying to avoid megadeaths from mass starvation and, quite possibly, from nuclear wars — and the odds aren't good.

This is a world in which food imports are no longer available at any price, as there is a global food shortage. But there are still relative winners and relative losers: the higher-latitude countries—northern Europe, Russia, Canada—are still getting adequate rainfall and are able to feed themselves, while those in the mid-latitudes are in serious trouble. Even the United States has lost a large amount of its crop-growing area as the rain fails to fall over the high plains west of the Mississippi, persistent droughts beset the southeast, and the rivers that provided irrigation water for the Central Valley of California cease to flow in the summertime. Countries of smaller size, like Spain, Italy and Turkey on the northern side of the Mediterranean (not to mention those on the southern side), find that their entire land area is turning into desert and that they can no longer feed their populations. The northeastern monsoon that brought rain to the north Chinese plain has failed, and the rivers that watered southern China have suffered the same fate as those that provided California's water: now they only flow in the wintertime.

This is a world where people are starting to starve, but it is not always the familiar scene of helpless peasant societies facing famine with numb resignation. Some of the victims now are fully developed, technologically competent countries, and their people will not watch their children starve so long as there is any recourse, however illegitimate, that might save them. So the lucky countries in the northern tier that can still feed themselves—but have little or no food to spare—must be able to turn back hordes of hungry refugees, quite probably by force. They must also be able to deal with neighbours who try to extort food by threats—and these desperate neighbours may even have nuclear weapons. Appeals to reason will be pointless, as it is reasonable for nations to do anything they can to avoid mass starvation.

If the climate modellers will not generate this kind of scenario, who will? The military, of course.

The military profession, especially in the long-established great powers, is deeply pessimistic about the likelihood that people and countries will behave well under stress. Professional officers are trained to think in terms of emergent threats, and this is as big a threat as you are going to find. Never mind what the pundits are telling the public about the perils of climate change; what are the military strategists telling their governments? That will tell us a great deal about the probable shape of the future, although it may not tell us anything that we want to hear.

In Britain, climate change has been taken seriously at the official level for a long time, and the British Armed Forces are free to discuss any scenarios they want. The *DCDC Global Strategic Trends Programme 2007–2036*, third edition, 2006, a ninety-one-page document produced by the Development, Concepts and Doctrine Centre within the British Ministry of Defence and regularly updated online, is "a source document for the development of UK Defence Policy."

In many ways, it is a remarkably sophisticated document. At one point, for example, it observes that "by the end of the period [2036] it is likely that the majority of the global population will find it difficult to 'turn the outside world off.' ICT [information and communication technology] is likely to be so pervasive that people are permanently connected to a network or two-way data stream with inherent challenges to civil liberties; being disconnected could be considered suspicious." But on the political and strategic impacts of climate change, it is surprisingly terse. Here is all it has to say on the matter:

> The future effects of climate change will stem from a more unstable process, involving sudden and possibly in some cases catastrophic changes. It is possible that the effects will be felt more rapidly and widely than anticipated, leading, for example, to an unexpected increase in extreme weather events, challenging the individual and collective capacity to respond . . .

Increasing demand and climate change are likely to place pressure on the supply of key staples, for example, a drastic depletion of fish stocks or a significantly reduced capacity to grow rice in SE Asia or wheat on the US plains. A succession of poor harvests may cause a major price spike, resulting in significant economic and political turbulence, as well as humanitarian crises of significant proportions and frequency . . .

Water stress will increase, with the risk that disputes over water will contribute significantly to tensions in already volatile regions, possibly triggering military action and population movements . . . Areas most at risk are in North Africa, the Middle East and Central Asia, including China whose growing problems of water scarcity and contamination may lead it to attempt to re-route the waters of rivers flowing into neighbouring India, such as the Brahmaputra . . .

A combination of resource pressure, climate change and the pursuit of economic advantage may stimulate rapid large-scale shifts in population. In particular, sub-Saharan populations will be drawn towards the Mediterranean, Europe and the Middle East, while in Southern Asia coastal inundation, environmental pressure on land and acute economic competition will affect large populations in Bangladesh and on the East coast of India. Similar effects may be felt in the major East Asian archipelagos, while low-lying islands may become uninhabitable.

There now, that wasn't so bad, was it? A shortage of fish here, a major price spike in food there, a little border war between China and India over re-routing the rivers, and a few tens of millions of climate refugees heading north out of sub-Saharan Africa and Bangladesh. If that's the sum of the damage climate change that will bring in the next thirty years, we can live with that.

Unfortunately, that isn't the end of
future-gazing only takes us out to 2036, not u
importantly, it is dated December 2006, which m
climate forecasts it is using come from the Intergoven.
Panel on Climate Change's 2001 report, not its 2007 rep
Essentially, the data it is using are on average close to ten years
old. That makes a big difference, because the data and the forecasts have been getting steadily worse. The next iteration of the
DCDC report will at least refer to the 2007 IPCC report
(although that is already seriously out of date, too), and is likely
to feature much darker scenarios on the climate-change front.

So, if the British Armed Forces aren't producing up-to-
date scenarios about the political and strategic impacts of climate change, who is? The American military? But here we have
the problem that the U.S. government, from the inauguration
of President George W. Bush in January 2001 until sometime in
late 2006, was in complete denial about climate change. In subsequent months the phrase "climate change" was finally heard
to pass the president's lips unaccompanied by disparaging
remarks several times, so, in late March 2007, the U.S. Army
War College sponsored a two-day conference on "The National
Security Implications of Climate Change," at which civilian
strategists and active duty and retired officers explored a wide
range of climate-related security issues. It seems clear that the
military had been chafing at the bit for some time previously,
however, since the following month saw the publication of a
study that had been in the works for at least two years. At the
time when it was commissioned, no bureaucratic warrior experienced in Washington's ways would have risked putting his or
her name on a study of the geopolitics of climate change, so
the Pentagon farmed the job out to the CNA Corporation.

I have long been interested in and concerned about
how environment affects security, and I spent eight
years at the Department of Defense with that portfolio,

environmental security. I was approached by a group of foundations several years ago and asked specifically if I would examine the national security implications of climate change, and for that purpose I assembled the Military Advisory Board of retired three- and four-star generals to assist us in that effort.

In our report, we were looking primarily over the next thirty to forty years. There are certainly disruptive events that could potentially occur earlier. An extreme weather event, or multiple extreme weather events, could occur at any time. But the more significant implications probably occur over the next several decades, and then of course far into the future. Unless we begin to reduce greenhouse-gas emissions and change the way we use energy, we really have some frightening futures.

—Sherri Goodman, general counsel, CNA Corporation,
in an interview with the author, February 4, 2008

The CNA Corporation is actually the old Center for Naval Analyses, descended from the group of scientists who brought the fledgling methodology of "operational research" to bear on the problem of anti-submarine warfare during the Second World War, and subsequently on other problems of naval strategy and tactics as well. It is now described as "a federally funded research and development center serving the Department of the Navy and other defense agencies." It produced its report, *National Security and Climate Change*, in April 2007.

The exercise involved choosing eleven recently retired three- and four-star generals and admirals from all four services, exposing them to the views of a large number of people working on climate change or related fields, and then writing a study on which the retired military men were asked to comment and elaborate. It created quite a stir when it was published, precisely because it effectively circumvented the Bush ban on treating climate change as a real and serious phenomenon.

You already have great tension over water [in the Middle East]. These are cultures often built around a single source of water. So any stresses on the rivers and aquifers can be a source of conflict. If you consider land loss, the Nile Delta region is the most fertile ground in Egypt. Any losses there [from a storm surge] could cause a real problem, again because the region is so fragile . . .

We will pay for this one way or another. We will pay to reduce greenhouse-gas emissions today, and we'll have to take an economic hit of some kind. Or we will pay the price later in military terms. And that will involve human lives. There will be a human toll. There is no way out of this that does not have real costs attached to it.

—General Anthony C. "Tony" Zinni, USMC (Ret.),
former commander in chief, U.S. Central Command,
National Security and Climate Change, April 2007

The *National Security and Climate Change* study is sixty-two pages long and very well sourced, but it doesn't really offer scenarios. It covers all the bad things that may happen if global warming progresses past a certain point, region by region, but it doesn't even specify what that point is. Indeed, it resembles a more concise version of all the books that have been published by various luminaries over the past couple of years rehearsing all the undesirable things that will happen to us if we don't pull our socks up and deal with global warming: a dab of science, a shopping list of small and large disasters in no particular order (not even in a likely time sequence), and a good deal of exhortation to take this seriously.

The real point of the exercise was probably to persuade a largely military audience of the importance of climate change by having the retired generals and admirals give it their imprimatur. A panel of experts wrote the actual report, but the senior officers were each given an entire page to express their views on the contents and the topic—and it is their testimony that is the heart of the matter. They are intelligent men of considerable experience, so they offer coherent and convincing testimony. But they are clearly selling something.

> People are saying they want to be convinced, perfectly. They want to know the climate science projections with 100 percent certainty. Well, we know a great deal, and even with that, there is still uncertainty. But the trend line is very clear. We never have 100 percent certainty. We never have it. If you wait till you have 100 percent certainty, something bad is going to happen on the battlefield. That's something we know. You have to act with incomplete information. You have to act based on the trend line . . .
>
>   The situation, for much of the Cold War, was stable. And the challenge was to keep it stable, to stop the catastrophic event from happening. We spent billions on that strategy. Climate change is exactly the opposite. We have a catastrophic event that appears to be inevitable. And the challenge is to stabilize things—to stabilize carbon in the atmosphere. Back then, the challenge was to stop a particular action. Now, the challenge is to inspire a particular action. We have to act if we're to avoid the worst effects.
>
> —General Gordon R. Sullivan, USA (Ret.),
> former chief of staff, U.S. Army,
> *National Security and Climate Change*, April 2007

What they are selling is a mission. The next mission of the U.S. Armed Forces is going to be the long struggle to maintain stability as climate change continually undermines it. The "war on terror" has more or less had its day, and besides, climate change is a real, full-spectrum challenge that may require everything from special forces to aircraft carriers. So it's time to jolt the rank and file of the officer corps out of their complacency, re-orient them towards the new threat, and get them moving.

Does this sound cynical? I don't really mean it to. The professional military exist because the civilian societies that pay for them believe they are necessary, and in a world of complexity and chance, where universal love has not yet been established as a governing principle, there are occasions when they are needed. It is their job to identify and define threats to the well-being of the society that employs them, and it is only as a by-product of that process that these threats also provide further justifications for the existence of the armies and navies. It took them a while, given the roadblock of the Bush administration, but they are definitely there now.

> MICHAEL KLARE: Not just the U.S. military but also the intelligence community . . . view climate change as a major factor in what the world will look like (in the future) and the consequences for national security, and they are deeply concerned about this.
>
> GD: What do you think made them shift?
>
> MICHAEL KLARE: Like everybody else, I think it's a change in consciousness. That's a combination of zeitgeist and the work of Albert Gore and the IPCC—everybody's consciousness has been changed by all of that. Number two, the scientific evidence has become overwhelming in the past couple of years, so they've been affected by that just like everybody else.
>
> GD: Is there also an element of opportunism here? The military always need threats in order to justify their budget. Is this a new one?
>
> MICHAEL KLARE: I would say that it's as much fatigue with their current mission as opportunism. Their current mission is Iraq and Afghanistan, and I know that the professional military is completely sickened and fatigued and exhausted with that mission, and I think that it must

be somewhat refreshing for them to talk about something that bears no taint whatsoever of the Bush administration, the Global War on Terror, Iraq, Afghanistan and
so on.

—Michael Klare, defense correspondent for *The Nation*,
in an interview with the author, March 9, 2008

Whatever their motives, the American military and intelligence communities are now fully committed to playing a
leading role in the struggle to contain the negative effects of
climate change. Indeed, there is some grumbling in
Washington that they are out to "militarize" climate change.
This new commitment is leading to the production, both
inside and outside the Pentagon, of serious studies of what the
future will look like politically and strategically as global
warming progresses, and what the role of the military will be
in that world. The most readily available of these studies is *The
Age of Consequences: The Foreign Policy and National Security
Implications of Global Climate Change*, co-published by the
Center for Strategic and International Studies (CSIS) and the
Center for a New American Security (CNAS) in November
2007. (As soon as it was completed, the team who wrote it was
asked to brief the National Intelligence Council.)

While CSIS is a long-established Washington think tank
with a broad range of interests, the Center for a New American
Security is a recent spinoff that focuses more directly on climate
change. Both institutions, however, are supervised by people
who have been at the heart of American debates on strategic
policy for decades. The board of trustees of CSIS includes former U.S. deputy secretary of state Richard Armitage, former
secretary of defense Harold Brown, former national security
adviser Zbigniew Brzezinski, former secretary of defense
William S. Cohen, former secretary of state Henry Kissinger,
former assistant secretary of state Joseph Nye, former secretary
of defense James Schlesinger, former national security adviser

General Brent Scowcroft, USAF (Ret.), and a who's who of corporate CEOs. The board of directors of the CNAS includes former secretary of defense William Perry, former secretary of state Madeleine Albright, former secretary of the navy Richard Danzig, former undersecretary of defense William Lynn, and General Greg Newbold, USMC (Ret.), former director of operations at the Joint Chiefs of Staff. It may also be relevant that the CNAS board of directors and the lead authors for *The Age of Consequences* include a significant number of former senior security officials in the Clinton administrations of 1993–2000.

The lead authors of the three scenarios in the study include John Podesta, who served as chief of staff to President Clinton in 1998–2000, Leon Fuerth, national security adviser to Vice President Gore and a member of the Principals' Committee of the National Security Council in 1993–2000, and R. James Woolsey, Jr., head of the Central Intelligence Agency 1993–95. Whether this might indicate that the study foreshadows the views of a possible Democratic administration post-2009 remains to be seen—and it should be noted that James Woolsey, one of the key participants, serves as a foreign policy adviser to the Republican presidential candidate, Senator John McCain.

The political/strategic scenarios elaborated by these authors are based on physical climate change scenarios developed from the data in the IPCC's 2007 report by Jay Gulledge, senior scientist and program manager for science and impacts at the Pew Center on Global Climate Change. The non-alarmist, "expected" scenario for 2040 begins with the A1B emission scenario in the IPCC's 2007 report, a scenario that assumes continued rapid economic growth in the emerging industrial powers like China and India, a mid-range estimate for human population growth, and significant advances in non-fossil-fuel energy technologies and in the efficiency with which fossil fuels are used. Of the six different scenarios that the IPCC considered, A1B is neither the most optimistic nor the most pessimistic, but it does assume a continuing widespread dependence on fossil fuels. Under this scenario, the atmospheric concentration of carbon dioxide will be nearing 700 parts per million by the end of the century, although by 2040 it probably will not have passed 500 ppm yet. (The pre-industrial concentration of carbon dioxide was 280 parts per million, and we are currently at 387 ppm.)

Most importantly, this first scenario in *The Age of Consequences* scenario accepts the IPCC's conservative assumptions about the "sensitivity" of the climate to increased levels of carbon dioxide in the atmosphere. It assumes that, by 2040, average global temperature has risen only 1.3 degrees Celsius above the 1990 average that the IPCC uses as a baseline, with a best estimate of 2.8 degrees Celsius above the 1990 figure by century's end. As non-alarmist scenarios go, however, it is already pretty worrisome, even for 2040.

The scenario goes something like this. Since temperatures are usually cooler over the oceans, which cover two-thirds of the Earth's surface, an average global temperature rise of 1.3 degrees Celsius would mean that it is 2 degrees Celsius hotter or more over the land masses, even hotter in the middle of the continents, and much hotter in the high latitudes—up to 4 or 5 degrees Celsius hotter in the high latitudes around the poles. Accelerated melting of glacial ice will raise sea levels worldwide by 0.23 metres by 2040 (with much more to come, of course), and that combined with more violent storm systems will produce storm surges that will inundate some densely populated river deltas, especially in South, Southeast and East Asia. Much land will be lost permanently, and tens of millions of refugees will seek new homes and livelihoods in neighbouring areas that are already fully occupied. Some of those areas will be across international frontiers, and the potential for conflict is very high. India, for example, is already building a 2.5-metre fence along the full length of its 3,000-kilometre border with Bangladesh, one of the countries that is likely to generate very large numbers of refugees as its low-lying coastal areas are lost to the sea.

Similar waves of refugees will be created in other parts of the world by massive droughts that drive farmers off their land, as global warming changes the rainfall patterns and deprives the subtropics and the lower mid-latitudes of much of their rain. There will be enormous pressures on the southern U.S. borders as Central America and the Caribbean reel under the combined impact of failing crops, more severe hurricanes, and sea-level rises. Europe's southern frontiers will face equal pressures from migrants from Africa—another early victim of failing rainfall—while the Mediterranean parts of the European Union will themselves be suffering from chronic and increasing drought. The southwestern United States will suffer more frequent and longer-lasting droughts that cause problems, not only for agriculture, but for its fast-growing cities, while low-lying coastal areas in the Gulf and mid-Atlantic states will face the risk of multiple Hurricane Katrinas. Some small island nations in the Indian and Pacific Oceans may have to be evacuated and abandoned altogether.

The near absence of tentative words like "would" and "may" in this section of the study is quite striking—but then, as the authors say, "It is not alarmist to say that this scenario is the best we can hope for. It is certainly the least we ought to prepare for." It is a deeply conservative forecast that presumes that no positive feedbacks kick in to accelerate the warming—and the authors find it so implausibly optimistic that they immediately offer an alternative scenario for 2040, which they entitle "Severe Climate Change":

> [This alternative scenario] assumes that the [IPCC 2007 report's] projections of both warming and attendant impacts are systematically biased low. Multiple lines of evidence support this assumption, and it is therefore important to consider from a risk perspective. For instance, the models used to project future warming either omit or do not account for uncertainty in potentially important

positive feedbacks that could amplify warming (e.g., release of greenhouse gases from thawing permafrost, reduced ocean and terrestrial CO_2 removal from the atmosphere), and there is some evidence that such feedbacks may already be occurring in response to the present warming trend. Hence, climate models may underestimate the degree of warming from a given amount of greenhouse gases emitted to the atmosphere from human activities alone. Additionally, recent observations of climate system responses to warming (e.g., changes in global ice cover, sea level rise, tropical storm activity) suggest that IPCC models underestimate the responsiveness of some aspects of the climate system to a given amount of warming. On these premises, the second scenario assumes that omitted positive feedbacks occur quickly and amplify warming strongly, and that the climate system components respond more strongly to warming than predicted. As a result, impacts accrue at twice the rate projected for emission scenario A1B.

And so, we are plunged into the nightmare world of scenario two, a world only thirty years hence in which the average global surface temperature is 2.6 degrees Celsius above 1990 levels, with higher temperatures over land and much higher temperatures in the high latitudes. Accelerated melting of the Greenland and West Antarctic Ice Sheets has already raised sea levels worldwide by half a metre, and storm surges driven by much more powerful weather systems are already causing crippling inundations in low-lying port cities like New York, Rotterdam, Bombay and Shanghai. London might buy itself fifty or a hundred years by building a second, higher Thames Barrier, but in general the outlook is for successive retreats inland to new, makeshift ports that will eventually be inundated in their turn as the sea level continues to rise. This continuing abandonment of existing assets and reinvestment in new, temporary port facilities will impose heavy burdens even on once-rich societies.

Meanwhile, densely populated river deltas, such as those in Bangladesh, Egypt and Vietnam, are already generating huge numbers of refugees as the land is eaten away by successive storm surges. Crop yields are falling steeply in these regions (which provide a disproportionate amount of the world's food). The irreversible destabilization of the ice sheets means that a further sea-level rise of four to six metres is inevitable over the next few centuries, so all the major river deltas are ultimately doomed, and civilization is condemned to centuries of continuous retreat as coastal lands are drowned.

Agriculture has become "essentially non-viable" in the dry subtropics as "irrigation becomes exceptionally difficult because of dwindling water supplies, and soil salination is exacerbated by more rapid evaporation of water from irrigated fields." Desertification is spreading in the lower mid-latitudes. Fisheries are damaged worldwide by coral bleaching, ocean acidification, and the substantial loss of coastal nursery wetlands—but then most major ocean fisheries will probably have collapsed through overfishing well before 2040 anyway, with no help from climate change. The scenario makes no attempt to calculate the global availability of food in 2040, but its many references to refugee flows and regional shortfalls indicate an implicit assumption that there is no longer enough food to go around.

But it is the magnification of these physical effects by likely political and social responses that particularly concerns the author of the "severe" scenario, Leon Fuerth. As he points out in the "Age of Consequences" document, "If the environment deteriorates beyond some critical point, natural systems that are adapted to it will break down. This applies also to social organization. Beyond a certain level climate change becomes a profound challenge to the foundations of the global industrial civilization that is the mark of our species."

Region by region, Fuerth assesses the probable impacts. In the United States, agriculture is practically at an end in California's Central Valley due to the failure of the rivers that used to be fed in the summer by the melting snowpack on the Sierra Nevada and Rocky Mountains, and the major cities of the southwest are suffering drastic, permanent water shortages. Rainfall declines steeply over the high plains west of the Mississippi, intensifying reliance on irrigation water pumped up from the giant Ogallala aquifer and speeding its depletion. Coastal populations in the southeastern states who are under constant attack from wild weather events will initially benefit from federal projects to protect them, but the attempts will fail: "The idea of resisting nature by brute engineering will give way to strategic withdrawal, combined with a rear guard action to protect the most valuable of our resources. Optimists might hope for a gradual relocation of investment and settlement from increasingly vulnerable coastal areas. After a certain point, however, sudden depopulation may occur." And under all these stresses, the author suggests, the federal system itself may start to weaken, with Washington offloading the burden of coping with the constant, multiple disasters onto state governments, as its own resources become inadequate for the task.

Meanwhile, the far more severe consequences of climate change in Mexico, Central America and the Caribbean, where drought has become the new normal, puts huge pressure on the U.S. border, where "problems will expand beyond the possibility of control, except by drastic methods and perhaps not even then. Efforts to choke off illegal immigration will have increasingly divisive repercussions on the domestic social and political structure of the United States." (By 2040—although the study does not explicitly mention it—some 20 percent of the U.S. population may be of Hispanic origin.)

Problems with Canada will accumulate, too, over fishing rights on both coasts, over water resources (especially if the U.S. decides to divert water from the Great Lakes, on which two-thirds of Canadians rely, to compensate for the effects of climate change elsewhere), and over navigation and resource rights in the newly ice-free Arctic Ocean. Moreover, the study states that "it cannot be excluded that Canada's tensions with the United States will play into domestic issues affecting the stability of Canada itself: most notably, the Western provinces' new role as oil exporter." (This is presumably a coy reference to separatism in oil-rich Alberta.)

In Latin America generally, the report predicts, severe climate change will be a death blow for democratic governments, and "Chavez-like governments will proliferate." Large regions will become essentially lawless or fall under the control of criminal cartels, and the United States, lacking the means to help local authorities to restore order, "will likely fall back on a combination of policies that add up to quarantine." The study implicitly assumes that the United States has already abandoned its more far-flung strategic commitments by 2040 and withdrawn to Fortress America, but it is a fortress surrounded by hostile neighbours. "The result . . . will be to render the United States profoundly isolated in the Western Hemisphere: blamed as a prime mover of global disaster; hated for measures it takes in self-protection."

Africa is the continent that takes the worst hit from climate change in almost every scenario, and this one is no exception. "The northern tier of African countries (i.e., Morocco, Algeria, Tunisia, Libya) will face collapse as water problems become unmanageable, particularly in combination with continued population growth," writes Fuerth, pointing out that, in Morocco's case, intense drought will destroy not only its irrigation agriculture but also its hydroelectric power generation. The countries of the Maghreb may try to tap into underground aquifers in a "zero-sum struggle for survival," but even the Great Nubian Sandstone Aquifer, currently the object of a US$20-billion Libyan mass irrigation project, would be largely drained in fifty years. Further east, wars between Egypt, Sudan and/or Ethiopia over attempts to divert the waters of the Nile and its tributaries for upstream irrigation projects are a growing possibility by 2040, and the whole Nile Delta is at risk from storm surges.

In sub-Saharan Africa, "hundreds of millions of already vulnerable persons will be exposed to intensified threat of death by disease, malnutrition and strife." The primary cause will be long-term drought, but the weakness of the infrastructure in most African countries will lead to a proliferation of failed states that exacerbates all the problems and generates huge waves of refugees. Many will follow the familiar paths north towards Europe, but there will also be a strong southward flow towards South Africa (which will be facing severe drought problems of its own).

In the Middle East, rapidly growing populations and declining water supplies will intensify existing hostilities everywhere. Attempts at an Israeli-Palestinian peace settlement will be abandoned indefinitely "because of a collective conclusion that the problem of sharing water supplies must be regarded as permanently intractable," and even war between Israel and Jordan over access to water is conceivable. Iraq, Syria and Turkey will become trapped in an "escalating struggle" over Turkey's control of the headwaters of the Tigris and Euphrates rivers. In the Gulf countries there will be a rapid expansion of nuclear power for desalination of sea water, and this will facilitate "the regional proliferation of nuclear weapons as insurance against predation."

All the Asian rivers that rise in the Himalayas and on the Tibetan plateau (Indus, Ganges, Brahmaputra, Salween, Mekong, Yangtze) will initially flood for decades as the glaciers and the snowpack melt, and then shrink drastically, especially in the summer months, once the glaciers and the snowpack are gone. This will lead to food shortages and cross-border disputes over water in the Indian subcontinent, and nuclear-armed India and Pakistan will face the risk of war over the Indus River. (The largest contiguous area of irrigated land on Earth is on the lower reaches of the Indus river system, in Pakistan, but the headwaters of the rivers are in India.) Indian democracy may fail under these stresses.

China's shrinking rivers will affect not only food production across southern China but also the country's ambitious hydroelectric schemes, like the Three Gorges Dam. The weakening of the northeast monsoon will cut grain production on the north Chinese plain, and China's industrialized coastal regions will take a severe battering from rising sea levels and stronger storm systems. The autocratic Chinese regime may seek to fortify its domestic position, rendered shaky by these blows, by directing popular anger outwards, at Taiwan, Japan or even the United States.

Authoritarian regimes are also likely to arise in Europe, especially in Russia, where the regime "will anchor itself ideologically in Russian nationalism, and economically on the basis of a dominant energy position, which it will exploit aggressively." But similar things will be happening politically in Western Europe under the impact of an influx of illegal immigrants from northern Africa and other parts of the continent. It will be an influx "impossible to stop, except by means approximating blockade." Hostility to Muslim communities in particular will increase; efforts to integrate them into the European mainstream will collapse, and "extreme division will become the norm." Economically, the European Union will have its hands full as almost every major port faces inundation and the whole country of the Netherlands has to be rescued from the sea.

Now, there are obvious criticisms that can be levelled at this scenario, and the most prominent ones arise from the American perspective of the author. Borders are not nearly as hard to control as he believes. Even the U.S. border with Mexico could be sealed, at a tiny fraction of the amount spent annually on the war in Iraq, if the United States ever decided that it was willing to forego the constant influx of cheap labour that is facilitated by the current, deliberately porous border controls. The notion that Europe cannot control its sea frontiers with Africa is simply laughable; it is just not yet willing to use physical force to defend them. And Fuerth's exaggerated concern about the reliability of Muslim minorities in Europe, though not without echoes in the debate in Europe itself, is primarily a reflection of post-9/11 American obsessions that do not address contemporary European realities.

Nevertheless, the true insight of Fuerth's analysis lies not in the regional analyses, but in his observation that "massive nonlinear events in the global environment will give rise to massive nonlinear societal events. The specific profile of these events will vary, but very high intensity will be the norm."

> LEON FUERTH: Complexity theory . . . originates in very obscure mathematics that were developed to try to describe extremely unruly physical events, but it has developed a path that runs towards human events as well, and especially towards the interaction of physical and human events, which makes it very interesting for the purposes of dealing with something like climate change.
>
> The essential insight in complexity theory would be: Don't think of this as a linear process. Think of it as a process where at any point some small change of inputs could produce a massive, unexpected flip in outputs . . . Expect that any solutions you apply are likely to further disturb the system, leading to an infinite series of surprises. Very different from the kind of approach that is often taken in public policy, which is that you only need to do THIS, and the problem will be solved now and forever . . . Once you realize that, you begin to try to analyze different regions of the world and even different countries from the perspective of how their political systems will change, whether these are domestic or international.

GD: What you're saying, essentially, is that we're looking at potential system collapse, politically as well as physically. LEON FUERTH: This whole thing is an interaction between human beings as a highly organized industrial civilization, and the world's physics and chemistry and so on, and the consequences of things that we already have done, and set in motion, before we were smart enough to recognize the patterns.

—Leon Fuerth, professor of international affairs,
George Washington University, one of the lead authors of
The Age of Consequences, in an interview with the author,
February 5, 2008

Among the non-linear political events Fuerth foresees in the event of severe climate change are class warfare "as the wealthiest members of every society pull away from the rest of the population;" an end to globalization and the onset of rapid economic decline owing to the collapse of financial and production systems that depend on integrated worldwide systems; and the collapse of alliance systems and multilateral institutions, including the United Nations. He suggests that massive social upheavals will be accompanied by intense religious and ideological turmoil, in which the principal winners will be authoritarian ideologies and brands of religion that reject scientific rationalism. Even more disturbing (and persuasive) is his observation that "governments with resources will be forced to engage in long, nightmarish episodes of triage: deciding what and who can be salvaged from engulfment by a disordered environment. The choices will need to be made primarily among the poorest, not just abroad but at home. We have already previewed the images, in the course of the organizational and spiritual unravelling that was Hurricane Katrina. At progressively more extreme levels, the decisions will be increasingly harsh: morally agonizing to those who must make and execute them— but in the end, morally deadening."

And so we come to the pandemics facilitated by the collapse of public-health systems in poor countries, the wars (including nuclear wars), and the other second-order consequences of the climate change scenario Fuerth was given. He suggests that we may see abrupt die-offs of the kind that have occurred on a smaller scale among ancient peoples under severe climate stress, and even pre-emptive desertion of urban civilization in some regions. Yet he does not come across as someone who is happy predicting doom; speaking to him, one has the impression that he was surprised, indeed shocked, by where his analysis led him. After all, the average global temperature was only 2.6 degrees Celsius hotter in the scenario Jay Gulledge gave him. (Although that does mean an average of 4 degrees hotter over land, and much more than that in the polar regions.)

In the end, I asked him how credible he thought his "severe" scenario was, compared to the "expected" scenario based on the most recent IPCC report. He replied: "The way more recent information is coming out suggests that the kind of future that's actually already loaded into the environment is better described by the kind of scenario that they gave me to work on—the 'severe' thirty years [scenario]—in which case the question is: what are we going to do to mitigate this so that we don't hand on to our [descendants] the severe one-hundred-year scenario, and what are we going to do to try to adapt to the consequences that may already be loaded into the system?"

In later chapters, I will deal with recent evidence that confirms Fuerth's sense that the severe thirty-year scenario is more credible. I'll also discuss what might be done by way of mitigation. But we should address the third one-hundred-year scenario now: "Catastrophic Climate Change." Written by former CIA head James Woolsey, this scenario assumes an average global temperature 5.6 degrees Celsius hotter than now and a sea level two metres higher (with much more to come). It contains all manner of blood-curdling predictions, such as a Sino-

Russian nuclear war in which China's desperate need to resettle tens or hundreds of millions of people driven from its
flooded coastlines leads it to try to seize a Siberia made more
agriculturally productive by the warming. But it is not as convincing a scenario as its predecessors, mainly because early
twenty-first-century American obsessions about Muslims in
general and terrorists in particular are transposed almost intact
into this scenario that is allegedly about the early twenty-second
century. A different kind of scenario for severe climate change
is called for at this distance from the present, and a persuasive
(though appalling) one is provided by the man who can fairly
claim to have been the founder of what we now call "Earth
System Science," the British geophysicist James Lovelock.

In Lovelock's most recent book, *The Revenge of Gaia*,
published in 2006, he observes that the concentration of carbon
dioxide in the atmosphere fell to 180 parts per million in the
depths of the last ice age, and rose to 280 parts per million after
it ended. The further rise to the current carbon dioxide level of
more than 380 parts per million is largely due to human activities since the beginning of the Industrial Revolution, so we
have already made as large a change in the composition of the
atmosphere as that which occurred between the last time when
glaciers covered much of the Northern Hemisphere and the
current warm spell.

The change in average global temperature between the
depths of the last major glaciation and the long interglacial we
now inhabit was about five degrees Celsius, so even assuming
that there are no nonlinear and unwelcome surprises (which of
course cannot be safely assumed) we may already be committed to an eventual further rise in average global temperature of
similar scale. There is no certainty about this, because we do
not know to what extent that earlier one-hundred-parts-per-
million rise in atmospheric carbon dioxide was supplemented
by various feedbacks in order to produce the five-degree-Celsius
rise in temperature between the last ice age and now. The extra

one hundred parts per million is already in the air (though it has only resulted in a 0.8-degree-Celsius rise in average global temperature so far), so we may already have blown it. But we cannot know that for certain, so it still makes sense to strive to curb our emissions.

However, as Lovelock points out, it is highly unlikely that we will be able to decarbonize the global economy fast enough to avoid a further rise in carbon dioxide concentration to five hundred parts per million or more. This is comparable to the level of carbon dioxide in the atmosphere at the time of the last really hot spell in the Earth's history, at the beginning of the Eocene era, fifty-five million years ago, the so-called Paleocene-Eocene Thermal Maximum (PETM). The world had been warming gradually for some millions of years, and reached a point at which warmer ocean water destabilized the clathrates buried beneath the ocean floor, especially in the North Atlantic. (Clathrates are deposits of methane, continuously pro-duced by bacteria in the deep ocean floor, that are contained in molecular "cages" of frozen water that are stable under the great pressures of those depths. However, the warmer the tem-perature, the more unstable the cages become and, at a certain point, they are liable to release the methane quite suddenly, resulting in enormous "burps" of methane gas that rise to the sea surface and thence into the atmosphere, where they are a powerful warming agent.)

The PETM episode was caused by the sudden release of between three hundred billion and three trillion tons of fossil carbon, probably mostly in the form of methane gas from the clathrate deposits in the North Atlantic. (Why the North Atlantic? Not clear, although at the time it may have been warmer than other oceans due to some vagary of the currents.) At any rate, the result was that the early Eocene world, which was already somewhat warmer than ours is today, with no ice at either pole, experienced a runaway heating of about six degrees Celsius over a period of only twenty thousand years. Most of the temperature change occurred in two thousand-year bursts at the beginning and end of that period, presumably corresponding to enormous clathrate releases at those times. The remarkable thing, however, is that there was not a mass extinction. There was a significant turnover in the mammal populations, with most of the primitive mammals that had developed since the end of the Cretaceous Period being replaced by the ancestors of modern mammal groups (all of them in small versions adapted to Eocene heat), but there was no actual reduction in the number of species. On the contrary, there was a major diversification of species in the subsequent hot period, when there were both trees and alligators in the polar regions, and the only major loss of species was in the deep ocean regions (again, principally, in the North Atlantic).

The disturbance lasted about two hundred thousand years, during which the lower and middle latitudes of the planet were largely barren of life: deserts predominated on land, and the upper layers of the oceans were effectively semi-deserts too, since the density of marine life plummets once the sea-surface temperature exceeds twenty degrees Celsius. Only in the higher latitudes around the poles were there reasonably temperate conditions in which land and ocean life could thrive—but it did thrive there, as twenty thousand years had given it enough time to migrate and adapt. Human agriculture and fossil fuel burning have already released five hundred billion tons of carbon, which places us within the range estimated for the Eocene event. If that is what we are about to unleash, however, neither we nor the rest of the planet's flora and fauna will have twenty thousand years to adapt: this time things are moving a lot faster, as James Lovelock writes in *The Revenge of Gaia*:

> The Earth has recovered from fevers like this [in the past] . . . but if we continue business as usual, our species may never again enjoy the lush and verdant world we had only a hundred years ago. What is most in danger is

civilization; humans are tough enough for breeding pairs to survive, and . . . in spite of the heat there will still be places on Earth that are pleasant enough by our standards; the survival of plants and animals through the Eocene confirms it . . . But if these huge changes do occur it seems likely that few of the teeming billions now alive will survive.

So, in Lovelock's hundred-year scenario—or two hundred years, or however long it takes for the full effect of the five-hundred-parts-per-million-plus carbon loading to be fully expressed in terms of higher global temperature—the great bulk of the Earth's land surface turns to desert and scrubland, with only the Arctic basin and Greenland remaining as "the future centres of an appropriately diminished civilization." With luck, a civilization of a few hundred million people might survive in this area, for "the tundra wastelands of Siberia and northern Canada that remain above sea level will be rich with vegetation, and the enlarged Arctic Ocean, filled with algae, may become the fishing grounds of the future." This is such a drastic scenario that I asked almost every climate scientist I interviewed whether it was over the top. Almost all of them took it seriously. In a February 2008 interview, Jay Gulledge:

> It's over the top only in the sense that scientifically we don't actually know what the consequences of our actions are for the Earth system. But everything [Lovelock] says has a functional basis, a theoretical basis. The most disturbing thing about the scenario that he develops is that it's plausible.
>
> A lot of people aren't used to thinking in terms of plausibility, and yet they do it any time they buy an insurance policy. One or two percent of people experience a fire at their house in their lifetimes, but all of them who have a mortgage have fire insurance. That's because it's

plausible that it could happen, and there's nothing in what Lovelock outlines that's unreasonable . . . The types of scenarios he draws are often dismissed because they seem so alarmist . . . but even though we don't know what's going to happen, what he says could happen. It's plausible.

RUSSIA, 2019

The biggest finding from our own work is not so much that the ice has retreated, which everybody pretty much knows, but that the ice that's remaining is a lot younger and a lot thinner than in the past . . . It takes less energy to melt thin ice, so the same amount of solar energy, the same amount of heat in the water, can melt that one-metre ice much quicker than it can the three-metre ice.

As we switch from a main ice pack that's two to three metres thick versus one metre thick, then we have the potential to lose more ice area quicker. Ice that's thinner typically only survives one melt season; ice that's thicker survives multiple melt seasons. Since about the late '80s, we've started to see what amounts to bites being taken out of this perennial ice . . . and the cumulative effect of that is that you started to have overall less surviving old ice.

Now this past summer we had first-year ice over more of the Arctic Basin than we ever saw before. The assumption might be that all that first-year ice will melt out this coming summer [2008], but now that you're having first-year ice further north than it typically was, you might expect that it's colder, so that ice might survive. So everybody's watching real carefully what's going to happen this coming summer, because if we see all that first-year ice melt out again, then probably we will have another record reduction in ice cover . . .

The climate models are suggesting that we really shouldn't be seeing these big changes like we had in 2007 until about 2030 or 2040, towards the midpoint of the century. If we see this a couple of years running, that tells us that it wasn't just a fluke; that we are about twenty or thirty years ahead of where we are supposed to be based on the climate models.

—James Maslanik, research associate,
Cooperative Institute for Research in Environmental
Sciences, University of Colorado at Boulder,
in an interview with the author, May 8, 2008

IT'S AMAZING HOW FAST IT ALL WENT WRONG. In 2005, the scientific consensus was that the ice cover on the Arctic Ocean was slowly melting, and that the Northwest Passage might be open to shipping by mid-century. In 2006 came the first few lonely scientists, looking at their data and climbing out on a limb, who suggested that the whole Arctic Ocean might be ice-free in late summer by 2013, and at that point, the petroleum geologists and the strategists got out their maps of the Arctic Basin. Then, in 2007, came the greatest publicity stunt of the decade, when Artur Chilingarov, the most famous Russian explorer but also a member of the Duma (parliament) and a confidant of then-President Vladimir Putin, took a submarine to the North Pole to plant a Russian flag on the seabed far below the ice. Then Chilingarov claimed it, like a sixteenth-century conquistador on some New World shore: "The Arctic is Russian. We must prove the North Pole is an extension of the Russian land mass."

ARTUR CHILINGAROV: This [Russian] flag is a copy of the flag which we put on the seabed of the ocean. I don't know why the Canadians reacted as they did.

GD: Was it your idea, the flag?

ARTUR CHILINGAROV: Yes, it was my idea. I'm a politician. I'm not only a polar researcher; my main job is politics, and everywhere I go I will raise up the Russian flag, whether it's the South Pole or a football match.

> —Artur Nikolayevich Chilingarov, polar explorer, deputy
> speaker of the Duma, Vladimir Putin's special envoy for the
> Arctic, in an interview with the author, April 23, 2008

The Canadians rose to the bait, with Prime Minister Stephen Harper paying a flying visit to the Arctic a week later to reassert Canadian control over the northern archipelago: "Canada has a choice when it comes to defending our sovereignty over the Arctic. We either use it or lose it. And make no mistake, this government intends to use it." He promised to build six to eight ice-strengthened armed patrol ships to guard the Northwest Passage, and a deep-water base for them on Baffin Island—although it was not clear whom the patrol ships were being armed against, since it was the Americans, not the Russians, who were at that time disputing Canada's right to control the Northwest Passage. All the nationalists applauded Harper's decisive action anyway.

The scientific evidence concerning the 2008 summer was ambiguous about the speed with which the ice cover on the Arctic Ocean would melt, so there were no more decisive actions for a bit, but a profound shift of perspective was taking place in all the countries that surrounded what was really a kind of cold Mediterranean. The Arctic Ocean was six times bigger, but it, too, was virtually surrounded by land. Hardly any of the hundreds of millions of people living in the countries around it had seen it in that light in the past, because it had no value for them—but now, in a matter of a few years, all that had changed. If the Arctic Ocean became open water, then the countries around it would be able to get at the fish in it, and the oil and gas under it. What you could only call a gold-rush mentality took hold.

It was typical of gold rushes in that there was unlikely to be nearly enough treasure to go round. A July, 2008 report by the United States Geological Survey estimated that the Arctic region might hold as much as 90 billion barrels of undiscovered oil. If true, that would amount to about one-eighth of the total oil that remained to be discovered on the entire planet (according to USGS estimates)—and the same report predicted that as much as a third of the world's undiscovered gas reserves were also to be found in the Arctic. But there were two statements in the report that should have undermined the gold-rush mentality and discouraged those who were tempted to make territorial grabs. The first was that the USGS's predictions were based not on actual seismic surveys but on a "probabilistic geological analysis": that is to say, they just counted up the geological formations that looked promising. The other, much more important caveat was that most of the suspected oil and gas reserves were under the continental shelves, quite close to the coasts of the countries that ringed the Arctic Ocean, and therefore already within their Exclusive Economic Zones. To a rational observer, that would have meant that there was no point in vying for a bigger share of the ocean bottom in the central Arctic Ocean: that was not where the oil and gas were likely to be found. Did that stop the nationalists in various countries from working themselves into a lather about it? Of course not.

As for Canada's Northwest Passage—or *not* Canada's Northwest Passage, depending on whom you believed—it was unlikely ever to become a major intercontinental shipping lane, even if it did save several thousand kilometres on the Panama Canal route between Europe and East Asia. The prevailing wind and current tended to push whatever ice there was into the channels between Canada's Arctic islands, so insurers would probably insist that only double-hulled ships sail through those waters. The melting of the polar ice and the transformation of the Arctic Ocean into open water was an ecological disaster, but it wasn't worth a tenth of the military attention that was lavished on it over the next couple of decades.

> Our confrontation is not with nature; our confrontation is with the United States of America. In that perspective, climate change brings both opportunities and dangers. Arctic melting is a real concern, and it is not in vain that our expedition was sent to the North Pole to probe the depths and leave a sign of Russian presence on the underwater mountains.
> —Alexander Dubin, former advisor to President Vladimir Putin, advisor to the speaker of the Duma, TV presenter, in an interview with the author, April 21, 2008

In the West, it is customary to blame Russia for what happened next, but that is just a local perspective. The reaction against the incompetence and corruption of the Russian "democrats" in the 1990s fuelled a resurgent Russian nationalism that managed to be nostalgic for Stalin and the tsars at the same time, but powerful nationalist forces were also at work in the other countries around the Arctic Basin. Take, for example, the 2011 agreement in which the U.S. recognized the Northwest Passage as sovereign Canadian waters, in return for full participation in the development of Canada's much larger share of the resources of the Arctic seabed. In order to gain privileged access to Canadian resources, Washington betrayed its prior agreement with the European Union to oppose Canada's Northwest Passage claim—and Canada sold out a big chunk of its resources in order to get American military backing for its much bigger seabed claim. Everybody was playing the same games, and the nature of the competition was driving them all in the same direction.

The basic problem was that there were no agreed boundaries to the Exclusive Economic Zones (EEZs) where the resources were thought to be. In May 2008, the five littoral countries—Russia, the United States, Canada, Denmark (via Greenland) and Norway—made an agreement to abide by the rules of the 1982 UN convention on the Law of the Sea, but there were already disputes over bits of territory, like the quarrel between Canada and Denmark over the ownership of tiny Hans Island at the top end of the Nares Strait, between northern Greenland and Ellesmere Island. There were self-serving disagreements over whether dividing lines between the EEZs should run perpendicular to the general trend of the coast at the border between two states, or along lines of longitude running straight north to the Pole. (The rules of this game are simple: each side chooses the principle that gives it more territory. In the case of the Norwegian-Russian dispute, this put 155,000 square kilometres of seabed rights in question.) And then there were the real deal killers: if the ice was melting fast, who was going to wait until 2020 for the scheduled decision of the UN authority that was to rule on the rival claims? And who would abide by that ruling if really valuable resources were at stake and some alternative justification for claiming them could be found?

The Law of the Sea said that countries were entitled to an Exclusive Economic Zone extending for two hundred nautical miles (about three hundred and seventy kilometres) from their coasts, but they also owned the seabed resources beneath the relatively shallow waters of their continental shelf out to 350 nautical miles—if the continental shelf extended that far. All the countries around the Arctic Ocean had until 2014 to get their claims in under this law (except the United States, which had not signed the treaty), and there was a great deal of scientific work going on, even under the ice, to identify seabed features which could be interpreted as extensions of the continental shelf. The Russians said that an underwater mountain range, the Lomonosov Ridge, that ran most of the way across the Arctic had its origins in their continental shelf, so their seabed rights went all the way to the North Pole. (That was what Artur

Chilingarov was really basing his claim on.) But a quite different principle had been applied in carving up the territory
around the other pole, in the Antarctic.

At the poles, all the lines of longitude converge, so between
1908 and 1942, seven countries with territory in the Southern
Hemisphere chopped Antarctica up into pie-shaped "sectors"
that all met at the South Pole. France, for example, owned some
uninhabited islands in the southern reaches of the Indian
Ocean, so it simply took the most westerly and most easterly
longitudes of those islands, extended them to the South Pole,
and claimed all the territory in between. A subsequent treaty
banning commercial exploitation of the continent (plus an ice
cap three kilometres deep) kept anybody from acting on their
claims in the Antarctic—but there were countries around the
Arctic Ocean that would do considerably better if the "sector
principle" was applied in the far north, too—notably Russia,
which would get half of the entire Arctic Basin. Indeed, the
Soviet Union had made a polar claim under exactly that principle back in 1924, and there was a good deal of suspicion that if
Moscow didn't get what it wanted under the Law of the Seas
rules, it would simply reject the ruling of the United Nations
tribunal and revive the old Soviet sectoral claim instead.

> We think the situation is very dangerous and serious,
> and we also think that NATO [North Atlantic Treaty
> Organization] will transform from a defence alliance to
> a bloc which will fight for energy resources, and it will
> fight for its interests by military means . . . Since
> 2002–03 the Norwegian Navy has had several warships
> protecting their fishing fleet off Spitsbergen, and I don't
> exclude that Russia might send its navy there, too.
>
> —Colonel Anatolii Tsyganok (Ret.),
> Centre for Military Forecasting,
> member of General Council of Russian Defence Ministry,
> in an interview with the author, Moscow, April 23, 2008

The event that tripped everybody into full military mode was the Spitsbergen Incident of 2012. Even allowing for the bad seamanship and the loss of Russian lives, it should have been just one more arrest by Norwegian patrol vessels of Russian fishing vessels trespassing in what Norway claimed as its EEZ: an official protest, an apology, and generous dollops of compensation. But the fish were swimming over disputed seabed that happened to be one of the three zones identified by the U. S. Geological Survey as most promising in terms of potential oil and gas reserves: the East Barents Basin. (The other two were the Alaska North Slope and the West Siberian Basin.) To make matters worse, the clash off Spitsbergen coincided with Dmitry Medvedev's campaign for re-election as president of Russia, despite Vladimir Putin's decision to take the job back, which pushed both candidates into harder and harder positions on the issue.

By the time Putin was sworn in as president in early 2013, the "Colder War" was a reality.

The United States was up for it. The immediate irritant was the fact that the Russian parliament had refused to ratify the agreement regulating the U.S.-Russian boundary in the Bering Straits for five years running (for once, the shoe was on the other foot, and it wasn't the U.S. Congress refusing to ratify a treaty), but the real problem was free-floating American anxiety about the gradual but unmistakable U.S. descent from the pinnacle of superpower dominance. The real challenger to the American position as sole arbiter of world affairs was China, but given the close trading relationship between those two countries, a military confrontation would have most unfortunate effects. Whereas Russia was not a major U.S. trading partner, the Russians were the familiar old enemy, and a lot of frustrated Americans were itching for a confrontation with some sort of foreigners.

The Western Europeans, on the other hand, would actually have preferred to sit this one out. After all, Norway, although a member of NATO, had neglected to join the real

European club, the European Union. However, the United States pressed them from one side, while their own newer EU partners in Eastern Europe, having lived for many decades under Soviet military occupation, were deeply and instinctively mistrustful of anything the Russians said or did. So, in the end, it was not just the Norwegians but all of NATO in a confrontation with Russia.

It was never quite as bad as the old Cold War. There were no tank armies stacked up on either side of the Russian border, and although the nuclear weapons were always there in the background, nobody really believed that they would ever be used. There was no genuine ideological quarrel, although both Russians and Americans were impelled by their own cultures to pretend that there was. In reality, it was like the competition that raged between the Great Powers over the division of Africa in the thirty years before the First World War (even down to the detail that the prize wasn't really worth the effort and the risk).

But nuclear weapons kept everybody reasonably careful, and for the most part, it was just an enormous waste of money and time: warships confronting one another as drill rigs followed the retreating ice to the edge of the continental shelf and beyond, fishing vessels being arrested and confiscated, quarrels over Russian gas deliveries to Europe and a frantic European search for alternative sources of supply. The biggest short-term cost was the damage that the confrontation did to free-ish global trade, as the rivals issued "with me or against me" ultimatums to smaller powers, forcing them to take sides. The long-term cost was the global deal on climate change, which had to wait more than a decade past its planned completion in 2012 because of the confrontation between the key players.

> Now NATO is not as dangerous as China. The most dangerous (for us) is China, and nobody talks about it . . .
> The Chinese central government says that it recognizes our borders, but the Chinese provinces still claim that

Chinese territory starts from the Urals [i.e., all of Siberia and the Russian Far East belongs to China] . . . Within fifteen years, China can reach the level of development of Russia, and then I do not exclude that conflict will start.

—Colonel Anatolii Tsyganok (Ret.),
Centre for Military Forecasting,
member of General Council of Russian Defence Ministry,
in an interview with the author, Moscow, April 23, 2008

What finally ended the Colder War was a combination of everybody getting disillusioned and the Russians getting scared: doubts were growing about the extent of the Arctic Basin's oil reserves (though they did find quite a lot of gas), and Russia was getting very anxious about China's long-term intentions. Which is not to say that China ever planned to invade Russia. It's just that, by 2029, the long-delayed consequences of the long failure to address the climate-change issue at a global level were finally becoming manifest, and the situation in China was desperate.

Drought was ruining crops across the north Chinese plain and failing rivers were having the same impact in the south. The storm surge of 2028, the worst of a bad decade, reached all the way upriver to Shanghai. The insurgencies in Tibet and Xinjiang seemed endless, and now were accompanied by terrorist attacks in China proper. Some of the richer provinces were taking their distance from Beijing, and the insecure post-Communist government certainly needed something to pull the population together. It found irredentist rhetoric about China's lost northern territories useful, though it showed no immediate signs of acting on its words. Nevertheless, the Russians got alarmed, and decided it was time to shut down the foolish Arctic quarrel with NATO.

Moscow went further than that, and secretly offered the NATO alliance a strategic partnership (as an eightieth birthday present, perhaps). Nobody said whom the partnership would be directed against, but everybody knew. However, the Western alliance decided not to make an enemy of China, which was no more than five years behind the United States in military technology. The Russians were on their own.

As it turned out, China's internal problems overwhelmed the state before anything could go wrong on the Sino-Russian border. The civil disorders of the 2030s—"civil war" is too strong a term—left the country essentially without a foreign policy, and the only problem that the Russian army faced along the country's extensive border with China was turning back the waves of refugees. It was very hard, though, because they were being sent back to almost certain starvation.

The real cost of this long and pointless confrontation was the lost time. Eight years had been wasted by the George W. Bush administration in the United States at the beginning of the century, and then, just as everybody was gearing up for a serious global attack on the problem of climate change, along came this ridiculous pseudo-war to block international cooperation and waste another twenty years. So far, we're getting away with it: the feedbacks still haven't kicked in. But then, they wouldn't have done so yet even if they were now inevitable. If we do get into runaway global warming ten or twenty years from now, we'll know who to blame. And much good that will do us.

CHAPTER TWO
An Inevitable Crisis

One of the things that struck me on my first day in space is that there is no blue sky. It's something that every human lives with on Earth, but when you're in space, you don't see it. It looks like there's nothing between you and the Earth. And out beyond that, it looks like midnight, with only deep black and stars.

But when you look at the Earth's horizon, you see an incredibly beautiful, but very, very thin line. You can see a tiny rainbow of colour. That thin line is our atmosphere. And the real fragility of our atmosphere is that there's so little of it.

—Vice Admiral Richard H. Truly, USN (Ret.), former
shuttle astronaut and commander of Naval Space Command,
National Security and Climate Change, April 2007

BUT THE SKY LOOKS PRETTY BIG FROM BELOW, and we feel very small beneath it. For this reason, it was very hard for previous generations to imagine that our own actions could affect the climate in any serious way. Even as our numbers and our powers grew, there was little to suggest that our activities were beginning to overwhelm the natural cycles that maintain the stability of the current global climate. Indeed, nobody knew that there had ever been different global climates in the past, some of them much less suitable for human mass civilizations to thrive in, nor did they understand that seemingly minor changes in the composition of the atmosphere could flip the climate from its present state of equilibrium to a quite different state in a relatively short period of time.

So nobody is to blame for the crisis that looms over us. Not my mother, who had five children and contributed to a population explosion that saw the world's population triple from 2.3 billion to 6.7 billion between the end of the Second World War and now. Not William Levitt, who invented the modern suburb in the late 1940s, or Henry Ford, who applied mass-production techniques to the manufacture of automobiles in 1913, or Thomas Newcomen, who devised the first practical steam engine in 1710. And certainly not the first woman who planted seeds with a digging stick ten or twelve thousand years ago and began the agricultural revolution that set us on the path to mass civilization. (In the Neolithic division of labour, women were the gatherers who dealt in plants, while men were the hunters.) None of them could imagine that we might actually change the way the world works to our own disadvantage.

Yet here we are, only six or seven thousand years after our ancestors built the first little cities, and the world is changing before our eyes. Runaway climate change threatens to sweep away our stable, familiar world and replace it with a terrifying chaos of famine, mass migration and war that could cut the human population to a fraction of its present numbers by the end of this century. And with the benefit of hindsight, we can now see that this crisis was inevitable from the moment that some people abandoned the hunting-and-gathering lifestyle and became farmers.

With a regular food supply that could be expanded simply by adding more farmers, the population of the farmers soared: any land that was even marginally suitable for agriculture could support between five and a hundred times as many farmers as hunter-gatherers. And the numbers counted: even though the hunter-gatherers were healthier, had more varied diets, and probably lived more interesting lives, when they came into conflict with the farmers over land, they usually lost. Within perhaps a thousand years, the farmers were a majority of mankind. Now they, and their urban dependants, account for 99.9 percent of the human race, and there are about a thousand times as many of us as there were in the days when everybody was a hunter-gatherer.

The other advantage of being a farmer was that you had a fixed address: you could own more than you could carry. And because the whole approach of farmers is to manipulate the physical environment, not just to cope with what nature provides, technology began to accumulate. Once writing was invented, knowledge started to accumulate at a higher rate, too. It took about ten thousand years, but, by the beginning of the eighteenth century, the combination of scientific knowledge and technological expertise achieved critical mass. If Thomas Newcomen had been struck dead by a meteorite in 1709, somebody else would have invented the steam engine—which unleashed the Industrial Revolution—within the next few decades. Indeed, it would probably have been somebody else in Britain, because that was where the critical mass of technology and scientific perspective came into being at that time.

The same applies to the invention of mass production— not actually by Henry Ford, but it was he who started mass-producing automobiles, which is relevant in the current context. The innovation of far-flung suburbs that made the sprawling metropolises of the present possible could not have come into existence without mass automobile ownership. One thing led to another with something close to inevitability: only the details of who was the actual innovator and when and where it happened were open to chance. So, here we are with about a quarter of the world's 6.7 billion people living in "post-industrial" but still high-consumption societies, and another half of the world's population going through a high-speed recapitulation of both the industrial *and* the consumption revolutions, and, of course, we are in trouble. We have grown very fast, we have appropriated a very large portion of the Earth's resources, and we may end up paying a very high price for it.

Yet nobody is to blame. We didn't realize—couldn't realize, given the state of our knowledge at the time—that our actions might affect the whole biosphere. Only in the past forty years have a few scientists begun to suspect that the climate might be changed by human activities. Only in the past thirty years have some people started to warn that the changes were actually underway, and only in the last twenty years has the science been good enough to pin the changes firmly on human activities. The crisis we face was foreordained from the moment that that first woman planted a seed, but it wasn't obvious to us. So we go into the crisis ill-prepared in both material and psychological terms, and the outcome is uncertain.

What we're seeing in the climate domain is not just increases in the average surface temperature of the Earth. People really need to understand that the average surface temperature is just an index of the state of the climate. It's sort of like the temperature of your body, and you say "What's a few degrees among friends?," and then you realize that if you have a fever of 40.5 degrees Celsius, even though that's only three and a half degrees above normal, it's potentially fatal. The same thing is true of the world: differences of a few degrees in the average temperature of the world reflect massive changes in the patterns of the climate that are determinative of human well-being. We depend on the climate for the productivity of farms and forests and fisheries, we depend on the climate for the availability of water . . . We are at risk from the climate from heat waves, from floods, from droughts, from wild fires, from sea-level rise. And what we are seeing is all of these things happening faster.

We are seeing not only a rise in the surface temperature of the planet. We are **seeing changes** in circulation patterns, changes in storm tracks, increases in flood intensity and frequency, increases in drought intensity and frequency, more and stronger heat waves, more powerful tropical storms—right across the board, everything that is expected to result from global climate change driven by greenhouse gases is not only happening, but it's happening faster than anybody expected.

> —John Holdren, director, Woods Hole Research Center,
> and past president, American Association for the
> Advancement of Science, in an interview with the author,
> February 6, 2008

It's the numbers that count, in the end. If there were still only the 250 million human beings who were alive at the time of the Roman empire, we could do almost anything we wanted with impunity: industrialize, eat lots of meat, drive big cars and fly halfway around the planet on holiday. Sooner or later, we would still have to address the issue of our greenhouse-gas emissions, for carbon dioxide stays in the atmosphere for about two hundred years, and even the emissions of a quarter-billion people living that way would eventually accumulate to the point where they had to start moving away from carbon-based fuels for their energy needs. But presuming that our science was good enough by then to figure out what was going on, a mere quarter-billion of us would have plenty of time to make the changes, and run little risk of facing an existential crisis along the way.

The billion people who were alive in 1800, at the very start of the Industrial Revolution, would have run into trouble a lot sooner even if they had kept their population under control, but they didn't. Human populations tend to grow up to the limit of their resources. Indeed, they frequently grow beyond them.

First, the Europeans at home and in their overseas colonies went through one of the greatest population booms in history, growing from 20 percent of the world's population in

1600 to about 35 percent by 1900. Most of their descendants (together with a few other people like the Japanese and the Koreans) live in developed societies, so about one billion people, the equivalent of the entire world's population in 1800, now consume food, burn fossil fuels and emit greenhouse gases at a fully industrialized rate. But second, and more importantly, the public-health measures that had allowed the population growth rates to soar in those countries soon spread to the rest of the world, triggering a comparable boom there, and so there are now more than five billion other people in the world as well. Over half of them live in countries where very high economic growth rates (6–10 percent annually) are now rapidly shifting the population towards late-industrial patterns of consumption and emission.

They cannot all get there: key resources (including oil and gas) are not sufficient to sustain three or four billion people in the current "developed world" lifestyle. Some scaled-down, heavily modified version of that lifestyle might be possible for such large numbers if it were attempted gradually and with great care over a long period of time, but the present headlong dash for growth will short-circuit the process by hastening the onset of an acute climate crisis long before the goal is reached. And the first and worst manifestation of that crisis will be in the world's food supply.

> [A rise of] one to two degrees would probably leave global food supply more or less okay. You would have some shifting (of food production) toward the higher-latitude countries and away from the countries that are close to the equator, where temperatures are already at critical thresholds. It's really the two to three to four degrees and more warming that can be particularly devastating. You can get there by the end of this century, and if you do get to that, then you're facing losses of output potential of something of the order of 20–25

percent in Africa and Latin America, 30 percent or more in India.

Although that's partially compensated by increased potential in some of the northern industrial countries, it would clearly be problematical. That's especially true if you don't have the benefit of what's called "carbon fertilisation": the fact that there's more carbon dioxide in the atmosphere, and carbon dioxide is an input into photosynthesis, so in principle you can get some offset from that . . . [But there is a] question of whether the laboratory tests, which do confirm higher yields, are realistic. Do they take account of open-air conditions? Do they take account of the need for other nutrients that go with the carbon? And basically the more recent experiments in the open air suggest that the earlier estimates had been exaggerated.

—William Cline, senior fellow, Peterson Institute for
International Economics, in an interview with the author,
January 31, 2008

All the other impacts of climate change—rising sea levels, bigger hurricanes and storm surges, the migration towards the poles of diseases now confined to the tropics—will arrive on schedule or before, but nothing matters as much to human beings as the food supply. Stop eating, and you will reduce your carbon footprint to zero in a matter of months. A higher concentration of carbon dioxide in the atmosphere probably has minimal positive effect on grain yields even in the temperate parts of the world, and all the other impacts of global warming on food production are bad.

Higher temperatures will have a disastrous impact on food crops in parts of the world where the average temperature during the growing season is already close to the maximum at which the plants can germinate. We have selected and cross-bred those plants over ten thousand years to be the best possible match for the existing climatic conditions, and they have very limited ranges of tolerance for different environments. While rice will germinate at temperatures as high as forty-five degrees Celsius, for example, most of the grains will be sterile if the temperature remains above thirty-six degrees Celsius for more than a few hours during anthesis, the most critical part of the growth cycle, and many tropical countries already experience extended periods during the growing season when the temperature is only a degree or two below that level. More generally, experiments in the Philippines suggest that rice yields fall by 15 percent with each one degree rise in average temperature during the growing season. We can (and should) launch crash programmes to breed new strains of key crops that are more tolerant of extreme heat, but we cannot be sure of success. We can be pretty sure that yields will be lower, even if we are successful.

A second problem is soil moisture. On average, there will be more rainfall on a warmer planet, since evaporation from the oceans will increase, but even areas that benefit from increased rain will also be suffering from higher rates of evaporation from the soil. As William Cline pointed out in the interview cited above, "the availability of soil moisture is the result of a race between faster evaporation as you get higher temperatures and, in some areas, at least, greater rainfall. The evaporation goes up very rapidly as you get higher temperatures, and so that race is basically lost in lots of countries that are closer to the equator, where there are already problems with the dry conditions." Places such as most of Africa, much of Central and South America, the entire Middle East, and a great deal of Asia.

So the tropics and the subtropics get hit first and worst, but it is generally assumed that temperate-zone agriculture will actually benefit from a little warming. Maybe increased productivity in the mid- and high latitudes will compensate for the fall in food production closer to the equator, or so it is hoped, but there are three unanswered questions here. One is whether this predicted increase in food production in the temperate latitudes could possibly be enough to cover the loss of tropical production *plus* the growing global population (another two-and-a-half billion by 2050?) *and* the rapidly rising per capita consumption of meat and dairy products (whose production eats up vast quantities of grain) in the fast-growing economies of Asia. A second is the question of how that hypothetical excess food, grown in the rich, high-cost economies of the developed world, actually makes its way into the mouths of the hungry poor in the tropics. Who's going to pay for it? And the third and biggest question is whether the prediction that food production will increase in the temperate zone, at least while we are still on the lower slopes of the global warming curve, is actually true.

Everybody agrees that, in a warmer world, with increased evaporation from the oceans, there will be more rainfall overall. The problem is that most of the rainfall may be in the wrong places. There will be plenty in the higher latitudes, but a good deal less in the mid-latitudes where we grow most of the world's grain. The breadbaskets of the planet—the American Midwest, the north Indian plain, the Australian wheatlands, the Mediterranean Basin, and so on—are likely to take big hits in terms of reduced rainfall and lower crop yields.

The reason for this is an atmospheric circulation pattern called the "Hadley cells." What drives this circulation pattern is the fact that warm, moist air is continuously rising at the equator—moist because there is high evaporation from the warm ocean surface, and rising because that is what warm air does. As it rises, however, it cools—the temperature drops three degrees Celsius with every thousand metres of altitude—and cool air cannot hold nearly as much water as warm air, so the moisture comes out in the form of tropical downpours. High above the equatorial regions, therefore, there is a constantly replenished layer of chilled, recently dried air—which is then pushed away, to both the north and the south, by more warm, moist air rising from below. This cold, dry air comes back down to the Earth's surface some 2,500 to 3,500 kilometres away from the equator, and as it descends it heats up due to the increasing pressure (a process known as "adiabatic heating"). When it hits the surface, it is both hot and dry. This is what causes the world's deserts.

The deserts are not randomly distributed around the planet. Most of them are arranged in two bands of desert girdling the planet north and south of the equator, precisely at the latitude where the Hadley cells bring their hot, dry air down onto the surface. (Then the circulation closes the loop by flowing back towards the equator on the surface—which is what gives us the "trade winds.")

Spin an old-fashioned globe mounted on a spindle and you will see the desert zones of the world blur into two yellow bands at around latitude twenty-five degrees North and latitude twenty-five degrees South. The Northern Hemisphere deserts caused by the Hadley cells, moving from west to east, are the Sahara Desert in Africa, the Arabian Desert in the Middle East, the Thar Desert in western India and southern Pakistan, and the Great Southwestern Desert in the United States and its Mexican counterpart. They are all along the same line of latitude. The similar band in the Southern Hemisphere (where there is much less land and much more sea) begins with the Kalahari Desert in southern Africa, continues through the great Australian Desert, and finishes up in the deserts of Peru and northern Chile.

Most of the world's richest breadbaskets, the places that are blessed with plenty of sunshine, long growing seasons and lots of rain, are located just a little further away from the equator than these deserts. Australia's wheat belt is immediately south of the desert all the way across the county from Perth to the Murray-Darling Basin. The traditional granaries of the Mediterranean countries and the Fertile Crescent are just north of the Saharan and Arabian deserts. The American Midwest is just north (and a bit to the east) of the great deserts of southwestern U.S. and northern Mexico. It would be disastrous if those desert bands expanded, encroaching on the breadbaskets—and that is exactly what will happen under almost any global heating scenario. Higher temperatures mean more energy in the system, and the Hadley cells expand, encroaching on the land that is now farmland. It doesn't all turn into desert, of course, but we can expect rainfall to drop by 25 percent, 50 percent, even 75 percent over the breadbaskets, depending on the specific region and the amount of heating that we are experiencing. That is not a happy thought, for we don't have much margin for error with the food supply.

Over the past sixty years, we have re-enacted the miracle of the loaves and fishes: we are now feeding three times as many people off roughly the same amount of land. In 1945, the world's population was just over two billion people, only double what it had been in 1800. In the ensuing sixty-odd years, it has more than tripled to 6.7 billion—and the vast majority of them have enough food to eat. Yet no more than 10 percent of the land we grow food on now was not already being farmed in 1945: human beings have been farmers for a very long time, and most promising farmland was brought under the plow a long time ago. Essentially, we have tripled the yield on the existing farmland.

This miracle owes something to the famous "green revolution," the biologically engineered new strains of familiar crops that were more drought-resistant, higher-yielding, more salt- and insect- and disease-resistant. But it owes more to brute force: in the postwar decades, we threw fossil fuels at the problem in a big way. Indeed, in a sense we are now eating fossil fuels: the amount of fertilizer we put on the land has increased more than tenfold since 1945, and the feedstock for nitrogen fertilizers is ammonia, which we obtain from natural gas. This shitload of fertilizer had dramatic results in terms of increased yield over the last half-century, but it is now hitting diminishing returns: not only does putting even more fertilizer on the land fail to raise yields as much as before, but even our current levels of fertilizer use are damaging the land and the water systems in a variety of ways.

We are also irrigating three times as much land as was irrigated in 1945. It still amounts to only 15 percent of the world's cropland, but it now accounts for 40 percent of total food-crop production. Did we discover a lot of new rivers after 1945? Of course not. Most of the new irrigated land depends on pumping water up from aquifers deep underground—in some cases a thousand metres underground. Once again, it is fossil fuels that drive the pumps that provide the water, and once again we are hitting diminishing returns. Many of the aquifers were filled millions of years ago and no longer have any natural connection with the surface, so they do not recharge: once all their water has been pumped up, that's the end of it. Others do refill gradually over time, but almost all of those are also being pumped at unsustainable rates.

In much of the world we also mechanized agriculture (more fossil fuels), although it is arguable that mechanized agriculture yields less per acre, on average, than the intensive cultivation and daily attention that a peasant farmer's family would bring to the same acre. At any rate, the six-and-a-half billion are being fed, if not in a very sustainable way. However, both the natural gas that makes the fertilizer and the oil that fuels the farm machinery and irrigation pumps are depleting as fast as the underground water. World grain production, which grew at an average of 2.5 percent annually through the half-century after 1950s, has now flatlined for some years. The population is still growing, and there are no more green revolution tricks left in the box. The recently revived claims that genetically modified crops will help to "fight world hunger" are as false and cynical as ever: GM food crops that have been engineered for drought resistance, salt tolerance, increased yield, etc., simply do not exist. Perhaps one day such wonder-crops will be developed, but all the commercially available

GM crops have simply been engineered to make them resist-
ant to the effects of certain patent herbicides and insecticides,
so that more of those products can be sold to farmers and
dumped on the land. We would have a major food supply
problem even if the climate remained stable.

It is true that birth rates have dropped in most countries,
below replacement level in many, but there is still a lot of population growth left in the system because human beings are
not salmon. We do not spawn and die. We have our children
and live on for many decades afterwards, so the net population
continues to grow for at least a generation and a half after the
birth rate drops below replacement level. For example, I and
my four brothers and sisters all married, and between us we
have ten children, so technically we dipped slightly below
replacement level (2.2 children per completed family) in our
own generation. However, we ourselves are all still alive many
years later, so there are twenty people where there once were
ten. Indeed, there are already six grandchildren, and still my
own generation lives on. This is happening worldwide, so the
global population is predicted to continue growing, at a diminishing rate, until it stabilizes at 8.5 or 9 billion people in the
latter half of this century.

But, it is extremely unlikely that there will ever be 9 billion
human beings on this planet. It's not just that there is no obvious
way to feed the next two-and-a-half billion. In the relatively
near future, global heating is going to start depriving us of a
large and steadily increasing portion of the food supply that
supports the present 6.7 billion. There will be famines, and a
great many people will die. So while we work frantically to get
our greenhouse gas emissions down to zero, we also have to
find ways of avoiding the wars that would increase the deaths
by an order of magnitude—wars that would also cripple our
attempts to avert the runaway climate change that would cause
megadeaths later on.

History shows, archaeology shows, that humans grow their population until they reach the carrying capacity of their environment . . . They've always done this: hunters and gatherers have done it, early farmers have done it, everybody in the world has done it. And when you reach the carrying capacity, in part because things are never constant . . . it doesn't take long before the climate gets you. The climate gets just a little bit worse, and suddenly you run out of resources, and you compete with your neighbours to get those resources so you don't starve. . . .

Humans have never been able to live comfortably within the carrying capacity. They've never been able to restrict their growth so that they stayed well below it, below it enough so that when the climate turns a little bit against them, it doesn't have a major effect. They've never been able to do this . . . The other thing that comes out of looking at the deep past is that this process can sometimes take a couple of hundred years. Your growth can be slow, the climate can get better and better for a while, you can sometimes let this go on for maybe two, maybe even three hundred years. I can't find any place in the world where it ever went longer than that without there being a crisis that resulted in warfare, depopulation, starvation, etc., . . .

If you want to get scared—we are now in a cycle that you could argue started with the Industrial Revolution in the early 1800s. We're now more than two hundred years into this, coming up on three hundred. We could be [just] one more example of the same phenomenon, but we think that somehow it's different this time . . . Since the 1800s our population has grown four- or five-fold, and it's quite conceivable that you could have a crash right back to where you were. When things are

good, the population grows, [and] we sow the seeds of
our own crisis.

—Stephen LeBlanc, director of collections,

Peabody Museum of Archaeology and Ethnology,

Harvard University, in an interview with the author,

February 8, 2008

We are clearly operating quite close to the limit of global
carrying capacity with existing food-production technologies,
but the rapid population growth of the past two centuries is
not inevitably destined to end in a population crash. There is
still a certain cushion of safety between us and too few calo-
ries per person worldwide (especially if we were to consume
more of the grain we grow directly, rather than feeding it to
animals to make meat). An end to the current population
surge is possible *without* the intervention of famine. It is just
not guaranteed.

On an optimistic reading of the statistical trends, the
world's population is heading for stabilization at eight or nine
times the pre-industrial level by the mid-to-late twenty-first
century. That is rather a lot of people: we and our domestic
animals would then account for more than half the total
weight of the land-dwelling mammals of the planet. But it is
probably within the capacity of current food production techniques to feed so many people, at least for some centuries, *provided* that this dense population was only faced by the sort of
modest, local cyclical fluctuation in climate that destroyed,
for example, the Anasazi societies of what is now the U.S.
Southwest in the thirteenth century. Climatic changes on that
historically familiar scale would not be global, they wouldn't
be permanent, and help from outside could sustain stricken
regions until the crisis passed.

The catch, however, is that this huge population boom
was only made possible by the exploitation of fossil fuels, whose
burning has produced as a by-product greenhouse gases that are

now changing the global climate on a scale at least an order of magnitude larger than the fluctuations of the past. Indeed, what we face now is not fluctuations at all, but rather the risk of accelerating, irreversible changes in the whole pattern of the world's climates—and most of those changes would have large negative impacts on food production. So maybe we don't escape from that three-hundred-year cycle after all—and the indisputable fact is that people (or at least people in the small-scale societies that anthropologists study) always attack the neighbours before they starve. Would that be true of large, developed societies too? How badly do we want to find out? Because even some relatively rich countries are going to have trouble feeding their people as global warming progresses, while some countries nearby will still have food. The pain will be unequally shared, at least in the early phases of the crisis, and it is this business of winners and losers that poses the greatest threat to global order.

Nearer to the equator, most countries will be in severe trouble: a recent study done by the Indian think tank Integrated Research and Action for Development (IRADe) concluded that a rise of only two degrees Celsius in global average temperature would cut Indian food production by 25 percent. Since India's one billion people grow just enough food to feed themselves now, that is the equivalent of around 250 million Indians with nothing to eat—and it is unlikely that they would be able to buy food from outside, since, in any global heating scenario, most regions of the world that now have a substantial food surplus will be hard hit as well.

India's food situation at two degrees Celsius hotter is desperate enough, but it does not begin to compare with the plight of Bangladesh, where the southern third of the country—home to sixty million people—would be literally disappearing beneath the waves due to sea-level rise. Relatively few people in Bangladesh understand yet why their country is vanishing from under their feet, but they will. They will also understand who caused this "climatic genocide," as Bangladeshi climate scientist Atiq Rahman calls it. Their bitterness will be very great.

> This is the ground zero of global warming. There is no question that this is being caused primarily by human action. This is way outside natural variation. If you really want people in the West to understand the effect they are having here, it's simple. From now on, we need to have a system where, for every 10,000 tonnes of carbon you emit, you have to take a Bangladeshi family to live with you. It is your responsibility.
>
> Atiq Rahman, executive director,
> Bangladesh Centre for Advanced Studies,
> interviewed in *The Independent*, June 20, 2008.

It is just ugly now, but later on it could get very dangerous across a lot of Asia. Pakistan, for example, whose vast area of irrigated land depends mainly on rivers that rise in India, is immensely vulnerable if India decides that its own needs come first. This is one part of the world where wars over water really are possible — and both India and Pakistan are already nuclear weapons powers. Bangladesh could be one, too, in a few years, if it really wanted to.

Africa will be the continent that takes the biggest hit from climate change, and it is already home to more than half the wars in the world. The impacts of climate change will probably trigger many more wars, but the brutal truth is that most conflicts in Africa do not affect the rest of the world. The Middle East is a very different case, but in this region climate-related changes are likely to exacerbate existing confrontations rather than create entirely new ones, so they need not be considered further at this point. The new and unfamiliar lines of potential fracture run between the northernmost tier of the major powers and the tier immediately to the south. Since these are developed countries, with very large industrial, technological and organizational resources at their command—and the ability to acquire nuclear weapons in fairly short order if they do not already possess them—it is the possible conflicts between them that pose the greatest threat.

In developed countries very far away from the equator, most farming areas will continue to receive adequate rainfall, and may even be a net beneficiary of the warmer temperatures. In particular, countries where the current northerly limit for agriculture is largely determined by the length of the growing season may find that the farming frontier has moved several hundred kilometres northwards as the number of consecutive frost-free days per year in those regions reaches the threshold needed for grain growing: as few as ninety days are enough in the really high northerly regions, where you can easily have sixteen hours of sunshine per day in the high-latitude summer. The big winners in this geographical lottery will probably be Canada, Scandinavia and, above all, Russia. (There are no comparable land masses at these latitudes in the Southern Hemisphere.)

In the other countries in the mid-latitudes of the Northern Hemisphere, it is a more complicated picture, but it generally resolves itself along a north-south axis. The more northerly tier of these countries, including the British Isles, most of France, the Low Countries, Germany, Scandinavia, Poland, Russia, Korea and Japan, and the densely populated parts of Canada, will continue to receive adequate rainfall, and these countries will generally be able to feed themselves. The more southerly countries, however, are likely to suffer severe declines in annual rainfall thanks to the expansion of the Hadley cells, resulting, in many cases, in recurrent or permanent drought. The affected regions would include Mexico, the Central American and Caribbean states, both sides of the Mediterranean, the entire Middle East, and the main grain-growing regions of Pakistan and India. Their main export, a generation hence, may be refugees.

Both of the world's greatest powers are also in the northern mid-latitudes, but their sheer size makes matters more complicated, for they straddle several of the zones mentioned above. For a variety of unrelated reasons, China is the big loser. While its most northerly region, Manchuria, comparable in latitude to the New England states in the United States, should continue to receive adequate rainfall at plus-two degrees Celsius, the north Chinese plain, where the country grows most of its wheat, will not: the monsoon that delivers the summer rainfall over most of that region is already beginning to fail, and in the longer run, the volume of flow in the Huang He (Yellow River) will be affected as well. To make matters worse, the shallow aquifer that underlies much of that region has already been pumped dry, and the deep aquifer is going fast. Farther south, where the main grain crop is rice, the glaciers and snowpack in the high Tibetan Plateau that feed the Yangtze and other major rivers are melting. When they are gone, the rivers will become more seasonal, and will contain less water in the summer months, when it is most needed.

> Northern Eurasian stability could . . . be substantially affected by China's need to resettle many tens, even hundreds of millions from its flooding southern coasts. China has never recognized many of the Czarist appropriations of Chinese territory, and Siberia may be more agriculturally productive after a 5 to 6 degree C rise in temperature, adding another attractive feature to a region rich in oil, gas and minerals. A small Russian population might have substantial difficulty preventing China from asserting control over much of Siberia and the Russian Far East. The probability of conflict between two destabilized nuclear powers would seem high.
>
> —R. James Woolsey, Jr., describing the security implications of his "catastrophic" one-hundred-year scenario in *The Age of Consequences*, November 2007

The United States is a harder case to call. It will probably suffer huge losses to its food production on the high plains west of the Mississippi River, where the rainfall will be diminishing at the same time as the giant Ogallala aquifer that provides irrigation water for the entire region is finally pumped dry. The Central Valley of California, which accounts for one-quarter of the food grown for human consumption in the United States, will face grave difficulties if the rivers that are fed by the snowpack on the mountains become seasonal (winter only): at two degrees Celsius hotter, much of the snow that now falls on the mountains in winter will fall as rain instead and run off immediately, leaving the rivers largely dry in summer. Nevertheless, the United States may still have enough good agricultural land in the "Old Northwest" and within a few hundred kilometres of the sea along the eastern seaboard, the Gulf coast, and in the Pacific Northwest to feed its own population, which is forecast to be four hundred million by 2050.

It is unlikely, however, that even the U.S. will still be in the food-exporting business once global heating reaches two degrees Celsius. Except for quite limited surpluses in Russia and Canada, nobody will be exporting food. If you cannot produce enough at home, then your people will just have to starve.

This is not just a formula for famine; it is also a formula for war. All around the world, countries facing mass starvation will be just a bit closer to the equator than countries that can still feed their people, and some of the countries on the losing side will be sufficiently developed to make their unhappiness with this outcome felt. (People always raid before they starve.) To make matters worse, it will not escape notice that the countries that suffer least from the changing climate are, in most cases, the ones that industrialized first, and that are responsible for most of the emissions that have set this lethal process in motion.

The outlook is not promising. The twenty years that have slid by with little action since NASA (National Aeronautics and Space Administration) scientist James Hansen gave his famous warning about climate change to the U.S. Congress in 1988 will not come back, and realistic people understand that it may be too late to avoid some truly unpleasant consequences no matter how hard we try to make up for the lost time. Concerted global action to cut emissions by, say, 80 percent in the next twenty years would avert the worst of the changes that lie in wait for us, but that would require the abandonment of "politics as usual." Moreover, the deeper we get into the food shortages attendant on global heating, the more difficult it will be to make international deals of any kind, though only global deals can save us. Once we hit mass famine, mass migrations, and widespread war, the game is lost, and the only rule is *sauve qui peut*. Every man for himself.

China is particularly worrying, as the country's insatiable need for energy imports, together with the general unsustainability of its present pattern of headlong economic growth (12 percent in 2007), mean that it may be heading for a crash: political and social upheavals may threaten internal stability, and paralyze the government's ability to make deals about greenhouse-gas reductions and carry them through. The Chinese Communist regime itself often warns that such upheavals could break the country's unity. That would leave behind squabbling fragments, with which it would be impossible for the rest of the world to make viable deals on curbing emissions.

But it is still too early in the game for despair. People and countries have in the past found ways to cooperate—not perfectly, perhaps, but well enough to avoid ultimate disaster. The most recent example (for which we are shamefully ungrateful) was the avoidance of a hemisphere-killing nuclear exchange during the forty years of the Cold War. Luck played some part in our escape, no doubt, but so did the dedicated efforts of several generations of people who, at the time, saw themselves on different sides of the barrier: those who opposed nuclear weapons altogether, and those who commanded them and kept

them in check. Most people are not stupid, and very few are suicidal. So, with this precedent in mind, what are the chances
that we can do it again—that is, make it all the way through the
twenty-first century without tumbling into runaway global heating and a massive dieback of the human population?

The country that has taken that question most seriously
is Germany, where the government has promised to cut
greenhouse-gas emissions by 40 percent by 2020, a target that
dwarfs anybody else's. (And Germany is, by the way, a cloudy
northern country with relatively little in the way of wind.) Dr.
Hans-Joachim Schellnhuber is the director of the Potsdam
Institute for Climate Impact Research and climate change
adviser to the German Chancellor, Angela Merkel. In an
interview in Potsdam on March 27, 2008, he explained:

> Here in Germany we feel that if we can confine global
> warming to two degrees Celsius above pre-industrial
> level—that's now the official goal of the European Union,
> and was heavily supported by Chancellor Merkel—then
> we would be in a position to avoid what we call "danger-
> ous" climatic change. We cannot be entirely sure. There
> are these tipping points out there (which is a special field
> of my own research), but we would at least have a fair
> chance to avoid the things which turn out to be unman-
> ageable. Two degrees warming would still mean a differ-
> ent world, but probably we would keep climate chaos at
> bay, at least.
>
> So let's focus on these two degrees first. In order to
> achieve that, we would have to reduce greenhouse-gas
> emissions by 2100 to almost zero. By 2050, we would have
> to reduce emissions at the very least by 50 percent glob-
> ally as compared to 1990. That would mean that the
> United States, for example, would have to reduce by 2050
> by 90 percent, Germany and the U.K. by 80 percent. If
> you follow that line, Germany with the 40 percent

reduction by 2020 would be on track, actually, but it would have to go into an 80 percent reduction by 2050.

Now there are some people, like Jim Hansen at NASA, who say we have to stabilize GHG [greenhouse-gas] concentration at 400 parts per million. We already have 382 parts per million, so it would be just 18 to go. I think 450 parts per million is—not a safe level, but a tolerable level. I'm concerned if people in the U.K. say that 550 parts per million is okay, because that would mean—best guess—at least three degrees of warming. Since the difference between an ice age and a warm age is just five degrees, that would be 60 percent of that distance, but we are in a warm age already, so it would mean that we would enter a hot age. That's not acceptable.

I think the 450 parts per million target, also known as the two degrees limit, is at least a good working hypothesis. If we go down this line of 40 percent cuts by 2020 and 80 percent by 2050, then we would develop an appetite for avoiding fossil fuels. Once we learn that we can change the energy system without becoming poorer, then I think we will do even more. We will become more and more ambitious as we go. That would be my strategy.

This 450 parts per million target—also known as the +2 degrees Celsius limit, although it only offers a fifty percent chance of staying below that temperature—has been widely adopted throughout the scientific community, and is now the official target of the European Community. Most of those who espoused it did so some years ago, basing their choice partly on the state of the science at that time but also calculating that that was the most ambitious target they could advocate without being treated as lunatics by the "practical" people who run our political and industrial systems. After all, we were already in the 380s, with high-speed global economic growth ADDING more than two parts per million to the total each year. How were you going to stop that juggernaut short of 450 parts per million? Indeed, how were you even going to stop it there? You have to be a realist. But even then James Hansen was saying that is would really be unwise to go beyond 400 parts per million. Now he's saying the safe limit is really more like 350 parts per million (which you can only see in the rear-view mirror).

The great uncertainty in all estimates of future climate change is how much temperature change will be caused by a given amount of change in the amount of carbon dioxide in the atmosphere. This is complicated by the fact that climate change is non-linear (e.g. the fact that the world warmed about 5 degrees Celsius between the depth of the Last Glacial Maximum 20,000 years ago and the onset of stable warm conditions about 12,000 years ago, while the carbon dioxide in the atmosphere increased from 180 parts per million to 280 parts per million, does not necessarily mean that the rise from 280 parts per million to 380 parts per million over the past two hundred years will bring a further warming of 5 degrees Celsius).

Moreover, whatever warming there is will be divided between fast-feedback climate effects occurring on a scale of years or decades (increased evaporation, changes in cloud cover, melting of polar sea ice, possible releases of methane from permafrost etc.), and slow-feedback effects like the melting of the glaciers on Greenland and Antarctica, followed by the spread of dark, sunlight-absorbing plant cover over large land areas previous covered with highly reflective glaciers, that operate on a timescale of decades to centuries or even longer. For these and other reasons, it has remained difficult to reach a consensus on exactly what level of carbon dioxide in the atmosphere creates a grave danger of irreversible warming big enough to take us right out of the Holocene, the interglacial period of stable climate in which human civilisation has grown and our number have increased almost a thousandfold. Decades of more and more sophisticated climate models have helped greatly, but the level of uncertainty remains—so in a recent scientific paper James Hansen and others took a quite different approach to the problem.

Paleoclimate data shows that climate sensitivity is approximately 3 degrees Celsius for doubled carbon dioxide, including only fast feedback processes . . . [I]ncluding slow feedback processes, [it] is approximately 6 degrees Celsius for doubled carbon dioxide for the range of climate states between glacial conditions and ice-free Antarctica. Decreasing carbon dioxide was the main cause of a cooling trend that began 50 million years ago, large scale glaciation occurring when carbon dioxide fell to [425 parts per million], a level that will be exceeded within decades, barring prompt policy changes. If humanity wishes to preserve a planet similar to that on which civilization developed . . . carbon dioxide will need to be reduced from its current 385 parts per million to at most 350 parts per million. . . . If the present overshoot of this target carbon dioxide is not brief, there is a possibility of seeding irreversible catastrophic effects.

Abstract of "Target Atmospheric Co_2: Where Should Humanity Aim?" by James Hansen, Makiko Sato, Pushker Kharecha, David Beerling, Valerie Masson-Delmotte, Mark Pagani, Maureen Raymo, Dana L. Royer and James C. Zachos, published at http://www.columbia.edu/~jehl/2008/target/co2-20080407.pdf

Rather than modelling today's and tomorrow's climate, Hansen and his colleagues looked at ancient climates and the transitions between them, to try to work out the relationship between changes in carbon dioxide concentrations and changes in average global temperature. Their first conclusion, drawing on reasonably reliable data for the recent Pleistocene era, was that the relationship between carbon dioxide and temperature is not all that non-linear after all, at least for the range of climates between deep glaciation (glaciers covering not only Antarctica but much of the northern hemisphere) and a much warmer, substantially ice-free world where the sea level is around 150 metres higher. At any point along that range of climates, doubling the amount of carbon dioxide in the atmosphere gets you 3 degrees Celsius of fast-feedback warming, followed by a further 3 degrees Celsius of slow-feedback warming. This is not happy news, as it means that there is no fortuitous non-linear escape hatch. The data from the deep past do not allow Hansen et al. to estimate how long it would take for the fast-feedback warming to occur in our current context, or how much longer it would take for the slow-feedback changes to produce the rest of the warming. In the deep past, the changes in carbon dioxide levels in the atmosphere, mostly driven by geological events like volcanism or weathering of rocks, were happening hundreds or thousands of times more slowly than the current anthropogenic changes, so we cannot look to the past for guidance on those questions. In particular, the ancient data give us no indication of the role, if any, that feedback mechanisms like rapid and massive releases of methane gas might play in determining the pace of the warming.

That's a pity, because this information may be of critical importance to us in the relatively near future: big methane releases could accelerate the warming well beyond what is predicted to happen from the accumulation of carbon dioxide alone. But methane releases are generally just an accelerant to global warming, not a fundamental factor, since methane lasts a relatively short time in the atmosphere. Ultimate outcomes are driven by carbon dioxide concentrations, and the key conclusion of Hansen and his colleagues was that if we double the carbon dioxide concentration in the atmosphere, we get the full 6 degrees Celsius of warming eventually.

The other conclusion of the Hansen team is about the carbon dioxide level that would initiate the irreversible melting of the Antarctic ice cap. There have been thick ice caps at the South Pole before, but so long ago that the data on what was happening in the atmosphere at the time when they finally melted is sparse to non-existent. However, we do have fairly reliable data for what was happening in the atmosphere when the current Antarctic ice cap began to grow in a previously ice-free world, some 35 million years ago. The concentration of carbon dioxide in the atmosphere at that time was 425 parts per million, with a margin of error of plus or minus 75 parts per million. If that was the point at which the Antarctic began to freeze when the carbon dioxide concentration was on the way down, it is a fairly safe bet that that is also the point when the last ice would melt in the Antarctic on the way back up — not right away, of course, but after all the slow-feedback warming has run its course. And that would be the last ice in the world: the Greenland ice cap and the glaciers of the high northern latitudes came later when the temperature and the carbon dioxide concentration were on their way down, so they would be gone earlier on the way back up.

These are numbers that we have to take with deadly seriousness. Allowing for that big margin of error (75 parts per million plus or minus), we can nevertheless be fairly sure that all the ice on the planet will eventually melt if the concentration of carbon dioxide in the atmosphere remains at some point between 350 parts per million and 500 parts per million for any lengthy period of time. We are already well past 350 parts per million, so there is a serious possibility that we are headed for a greenhouse planet even if we stop the increase right now — which is, of course, impossible. Hundreds of millions would starve within a year. We'll be very lucky if we can stop the rise before we reach the 450 parts per million level that the European Union and some other organisations have adopted as their "never-exceed" ceiling, but that level gives a very high probability that all the ice eventually melts. Therefore, James Hansen and his colleagues now argue that our target should be not 450 parts per million, but 350 parts per million. Then we would probably be safe. According to Hansen:

You know, the logical thing, if you want to keep the climate similar to that of the Holocene, you've probably got to keep the carbon dioxide not too different than what it was in the Holocene, which means on the order of 300 parts per million. . . . To figure out the optimum is going to take a while, but the fundamental thing about the 350 [parts per million target], and the reason that it completely changes the ball-game, is that it's less than we have now. Even if the optimum turns out to be 325 or 300 or something else, we've go to go through 350 to get there. So we know the direction now that we've got to go, and it's fundamentally different. It means that we really have to start to act almost immediately.

Even if we cut off the coal emissions . . . carbon dioxide would get up to at least 400, maybe 425, and then we're going to have to draw it down, and we're almost certainly going to have to do it within decades. . . . So how can you draw down the carbon dioxide, and how much can you draw it down?

We have a pretty good idea. You can sequester carbon in forests and in the soil. In both cases we're doing a bad job [at the moment]. We're actually losing carbon to the atmosphere from the soil and from the forests we cut down. We could improve our practices. In the case of agriculture, you can work bio-char, charcoal, into the soil, and use no-till agricultural practices in more areas where it would work well, and that allows carbon to be stored in the soil instead of escaping to the atmosphere. And we could re-forest areas where there's degraded land now that's not being used even for agriculture; we could store a lot of carbon in those trees. If you look at this quantitatively, you could probably store about 50 parts per million of carbon dioxide that way. But if we just continue to burn coal, we're going to be at 450, 500, 600 parts per million or larger, and then I don't see any way you can solve it by means of these fairly natural solutions of trying to re-forest and improve the productivity of the soil.

—James Hansen, director, NASA Goddard Space Studies
Centre, in an interview with the author, June 28, 2008

There is no need to despair. The slow-feedback effects take a long time to work their way through the climate system, and if we could manage to get the carbon dioxide concentration back down to a safe level before they have run their course, they might be stopped in their tracks. As Hansen et al. put it in their paper:

A point of no return can be avoided, even if the tipping level [which puts us on course for an ice-free world] is temporarily exceeded. Ocean and ice-sheet inertia permit overshoot, provided the [concentration of carbon dioxide] is returned below the tipping level before initiating irreversible dynamic change. . . . However, if overshoot is in place for centuries, the thermal perturbation will so penetrate the ocean that recovery without dramatic effects, such as ice-sheet disintegration, becomes unlikely.

The real, long-term target is 350 parts per million or lower, if we want the Holocene to last into the indefinite future, but for the remainder of this book I am going to revert to the 450 parts per million ceiling that has become common currency among most of those who are involved in climate change issues. If we manage to stop the rise in the carbon dioxide concentration at or not far beyond that figure, then we must immediately begin the equally urgent and arduous task of getting it back down to a much lower level that is safe for the long term, but one step at a time will have to suffice. I suspect that few now alive will see the day when we seriously start work on bringing the concentration back down to 350, so let us focus here on how to stop it rising past 450.

One hopeful recent development is that the U.S. foreign-policy establishment and the military are both now fully alert to the threat posed by climate change: the United States will almost certainly cease to be an obstacle and turn into a leader in the attempt to get global warming under control once a new administration takes office in 2009. Will this be enough to transform the global political dynamics and make a comprehensive political agreement on curbing greenhouse-gas emissions possible? It might be, but only if it were just the old industrial nations who had to curb their emissions drastically in order to avoid runaway climate change. However, they are not alone in the world, and will not long remain the only industrialized countries.

Most of the older industrialized countries have not increased their emissions significantly since 1990. Their current pattern of energy use is so profligate that they could cut consumption by at least 20 percent without suffering dramatic changes in their lifestyles. With greater effort and a little sacrifice, conservation could cut their energy requirements by 30 percent, and their greenhouse-gas emissions by almost as much. But cuts in greenhouse-gas emissions almost three times that deep are needed by 2050 if we are to avoid soaring past 450 parts per million of carbon dioxide, which means that not only the demand side of the equation but also the supply side must be changed. That is much harder to do.

This is an energy-intensive civilization built mainly on the availability of plentiful and cheap fossil fuels, and it cannot sustain anything like its present numbers without continuing to use enormous amounts of energy. Switching from fossil fuels to non-carbon sources for most of that energy is theoretically possible, but doing it in a relatively short period of time, without ever interrupting the supply, is akin to changing the engine, the driveshaft and all four wheels of a moving car without ever stopping it.

There is also the political problem that the alternative-energy sources may have to include a large-scale return to nuclear-power generation. Hydro is good, but, in most countries, most of the good sites have already been developed. Tidal power is promising for some countries, but is still in development. Wind and waves and sunshine are all excellent carbon-free sources of power, but there are times when the wind does not blow and the sun does not shine. To generate the "base load" of electricity that is always on, it is argued, there is a need for a large, non-fluctuating source of power, and the only substitute for fossil fuels that is technologically mature and immediately available may be nuclear power. Even to raise this issue causes fury in a large section of the environmental community, but there is going to be a political struggle over increased reliance on nuclear power that will split the green movement and perhaps the general public in every Western country, and may significantly delay the adoption of either that or any alternative solution to the problem.

If agreement on alternatives can be achieved fast enough, the developed countries will probably be able to replace most of their power-generating plants with non-fossil sources in as little as two decades. Combined with conservation measures, that would achieve the emission cuts that most climate-change experts deem necessary, if we are to avoid disastrous changes later in the century. After all, the design lifetime of most power plants we build now is forty or fifty years, so we are, in practice, committed to replacing our entire power-generating capacity with new plants every half-century anyway. Doing it in twenty years is obviously a big step up from normal service, but it is *not* an unimaginably great change. If the demand is present, so will be the supply.

Unfortunately, it is not just the older industrial countries that must curb their emissions. Most of the growth in emissions at present is actually in the newly industrializing countries of Asia and Latin America. China's emissions are growing so fast that some experts believe it overtook the United States to become the world's biggest emitter in 2007, and countries like

India, Brazil and Mexico are following the same path, although not on the same scale.

These are still relatively poor countries, and so their best option for keeping the lights on, as the demand for electricity grows by leaps and bounds, is the cheap and dirty solution that everybody in Europe also chose before they got rich: coal-fired generating stations. (China is currently opening a new, large, coal-fired power plant *each week*.) These are countries that can practically taste the prosperity that only one or two more decades of high-speed industrialization will bring, and after centuries of being poor and disrespected they have no intention of stopping. Yet it will do little good for the older industrial countries to curb their greenhouse-gas emissions if the newer ones let their emissions rip. Some deal must be devised that will let the emerging countries continue to grow their economies without tipping the whole world into catastrophic climate change.

The outlines of the deal have been obvious for ten or fifteen years, but the political obstacles are huge. (So huge that no attempt was made to include the emerging economies in the emission caps and reductions mandated under the much-abused Kyoto Accord in 1997.) To suggest that the developing economic giants accept the same curbs as the fully developed countries while there is still such a great gulf between the living standards of their citizens is simply to invite a punch in the face, so the deal has to include two key elements. First, the rich countries have to accept even deeper cuts in their emissions in order to leave the emerging economies some scope for expanding their emissions. Second, there has to be not only technology transfer, but direct financial subsidies from the developed countries on an extremely large scale, in order to give the developing countries adequate resources for the task of switching their power-generation capacity from fossil sources to (more expensive) non-fossil technologies. Governments on both sides of the fence understand the shape of the deal that must be done, but the politics of it is very hard to manage in the developed countries, while the managers of electricity grids in the fast-growing countries of Asia are in a race just to keep the lights on. They will need a lot of help if they are to make the right choices.

Negotiating such a deal—a global deal, with everybody included—will be the second-hardest political enterprise ever undertaken. The hardest will be selling the completed package to the political audiences back home. The underlying principle that would shape this deal (if it happens) and justify the vast transfer of resources is that everybody on the planet is entitled to the same basic personal allocation of greenhouse-gas emission rights, and that those who exceed that allocation must compensate for those who use less than their permitted amount. If that deal cannot be made, then we must live with the consequences. Or die from them.

SCENARIO THREE:
UNITED STATES, 2029

It's conceivable that at two degrees Celsius there may be no significant change globally [in food production], but a major change in distribution. That is to say, there could be an increase in agricultural productivity in the mid- and high latitudes of the Northern Hemisphere, 'cause you'd have a longer growing season and hence more productivity. It's probably not until you get to maybe two-and-a-half, three degrees Celsius, that it would start to become a more negative effect. But clearly *any* warming—one degree, two degrees—would have a very significant adverse effect throughout most of the tropics and most of the subtropics.

—Robert Watson, chief scientific adviser, Department of Environment, Food and Rural Affairs, London, and former chair of the Intergovernmental Panel on Climate Change, in an interview with the author, May 19, 2008

TWENTY YEARS AGO, there was just an ordinary chain-link fence stretching dead-straight across the Arizona desert from horizon to horizon. Its purpose would not have been immediately clear, because it was much too elaborate for cattle (who couldn't survive in this environment anyway) and there was nothing of obvious value on either side of it. But one side was Mexico, of course, and the other side was the United States, so there were lots of people who wanted to get from one side to the other. The purpose of the fence had been to make it hard—but not impossible—for illegal immigrants to get into the United States. It was, in that sense, essentially a cosmetic device, although sometimes people died because of it.

The chain-link fence or its lineal descendant is still there, as easy to climb over or cut through as ever. However, it now bears warnings in Spanish and English to proceed no further on danger of death, and the signs are brightly lit at night. About two hundred metres to the north, on the American side, there is a much more serious barrier: two parallel open-mesh fences, three metres high, with razor wire on top, and separated by a raked sand strip, fifty metres wide, and a dry moat, three metres deep.

There are closed-circuit television cameras atop these fences (including ultraviolet ones for night vision), and movement sensors buried beneath the sand strip, and anti-personnel land mines in the dry moat. There are also automated machine guns atop the northernmost fence every four hundred metres, all the way from the Gulf of Mexico to San Diego on the Pacific Ocean. All that talk about how you can't seal off the Mexican border was just deliberately misleading propaganda from people who wanted to keep the border porous. If the Russians could build an Iron Curtain across the heart of Europe in the twentieth century, then the United States can do the same thing along the Mexican border in the twenty-first century. Except that this barrier is to keep people out.

Back at the beginning of the twenty-first century, a chain-link fence had been all that was needed. Only a couple of million Mexicans and Central Americans tried to cross into the United States each year, and American agribusiness needed at least half of them to make it through in order to replenish the supply of cheap illegal labour that made the farms profitable. It was all done on a nod and a wink, with Congress voting substantial sums annually to guard the border, but not passing any law that seriously penalized American employers for hiring illegal immigrants. The border had to be sufficiently guarded to satisfy general American public opinion, but not too well guarded for fear of choking off the supply of cheap labour.

So, gradually, all the easy crossing points in and near towns were shut down (they were too visible), while out in the open desert there was still only the chain-link fence. It was patrolled, of course, and lots of the people trying to cross it were arrested and deported back to Mexico (from where, unless they had run out of money to pay the "coyotes" who guided them through the desert, they would try to cross again a few days later). But many people did make it across—and every year some hundreds lost their way and died of thirst and heat in the baking desert. It looked better that way.

Time passed, and things changed. By the mid-2020s, the permanent drought that had long been forecast as the fate of the subtropics when global average temperature rose beyond one degree Celsius was turning the farms to dust from Mexico to Costa Rica. With no agricultural hinterland left to service, the towns were emptying out, too. Tens of millions of people were on the move, many of them piling up in the big Mexican cities (which were breaking down under the strain), but many more pushing on to the U.S. border. With over a million people trying to cross the border each month, the old catch-some-and-let-some-go system broke down completely. The existing fences along the border were a joke: the coyotes found a hole, cut a hole or blew a hole in them—and if U.S. border security agents intervened, they sometimes blew holes in them, too. By 2026, about half a million illegal immigrants were streaming into the United States each month. Social services in the border states were collapsing under the pressure of providing basic emergency services to the newcomers, and American public opinion finally demanded that something serious be done about the border.

It was done. In 2029, the frontier fortifications stretched three thousand kilometres from the mouth of the Rio Grande in Texas to the suburbs of San Diego in California, and the border truly was sealed. After some very ugly incidents early on when groups of would-be immigrants tried to cross the completed sections of the "Big Fence" and were practically wiped out by the automated weapons and mines, attempts to sneak across the border in the old style practically ceased. There was no more illegal immigration from Mexico to the United States (and very little immigration of any sort). A less visible but equally lethal barrier of patrol ships and aircraft in the Gulf of Mexico and around the Florida coasts stopped climate refugees from Cuba, Haiti, the Dominican Republic and the smaller Caribbean Islands, all of which were experiencing the same severe drying conditions and devastating hurricanes. The total monetary cost of the land-border fortifications and their seaward extension was US$1.8 trillion (about US$265 billion in 2016 dollars, before the Great Inflation). The human and political cost was far greater.

> The fact that our government in the 1980s opened our borders to imports of corn and other basic products from the U.S. was really a mistake. I kept saying that as a Mexican and as a social scientist, and now I am very unhappy to see that I was right. The food sector didn't receive the support they needed to compete in a market which is subsidized. Let's be honest, Europe and the U.S. and Japan subsidize their agriculture, and we didn't do it, and then we came to compete with those guys, the result being that farmers were forced to migrate to cities and the U.S.
>
> This will be aggravated by climate change . . . Mexico is already facing lots of stresses—social, economic, open market, etc. Okay, people are already migrating because of those reasons. Add to that climate change, and you have a bomb. Yes, the U.S. should be worried—but not only [respond by closing the] borders.
>
> —Paty Romero-Lankao, deputy director,
> Institute for the Study of Society and Environment,
> National Center for Atmospheric Research,
> in an interview with the author, May 7, 2008

For the Mexican state as it had existed for the previous hundred and twenty years, the Big Fence was a death blow. As late as 2020, Mexico was the twelfth largest economy in the world and a full member of the North American Free Trade Agreement (NAFTA), although the various peasant insurgencies in its southernmost provinces, intensified by the unprecedented drought that had struck the region, meant that it had lost control of significant tracts of its own territory. Over the following years, as the northward exodus from the Central American countries and southern Mexico gathered momentum, vast squatter camps of climate refugees built up around all the big cities of central and northern Mexico, and in some of the camps—especially those with large numbers of people from Guatemala, El Salvador and Oaxaca and Chiapas states (where the peasant insurgencies were most powerful)—armed militias gradually took full control.

By 2026, there were large areas around every Mexican city that had become no-go zones for the police and even for the army, and malnutrition had reappeared as a major problem among the poor. Nevertheless, the elected authorities were able to retain overall control thanks to the relatively open U.S. border, which served as a safety valve for the truly desperate. The construction of the Big Fence closed the safety valve.

In retrospect, everybody seems to agree that the Jimenez government's decision in 2028 to withdraw from the NAFTA treaty in protest against the closing of the border was a fatal blunder. By withdrawing, Mexico also effectively cancelled the Food Protocol to the NAFTA agreement, negotiated in 2014 when the world grain markets were first seizing up, and so lost access by treaty right to American and Canadian grain at the domestic market price. Mexico could not afford to buy grain on what remained of the global free market (most of the remaining international grain trade was by now tied up in long-term bilateral deals), and the United States and Canada, two of the very few countries that were still able to export large quantities of grain, simply sold it to the highest bidder instead—usually China. But the critics are being unfair: Jimenez had no choice but to cancel the NAFTA treaty in the face of popular fury at the video images of Mexican families being machine-gunned by American robot weapons at the border.

Nevertheless, the decision was fatal for the Mexican state. Deprived of access to relatively cheap NAFTA grain, the government's ability to distribute subsidized emergency food to the squatter camps and Mexico's own growing army of the utterly destitute rapidly dwindled away, and its authority drained away with it. By early 2029, some of the more powerful state governors were usurping federal powers, including control over local army units, and began to hoard locally produced food for local consumption.

The food crisis in Mexico City, whose population had swollen to twenty-seven million people (including refugees), now turned into a general health crisis, with malnourished people falling victim to epidemics in large numbers. Big groups of squatters and refugees fled the disease-ravaged slums and began to roam the countryside all the way down to the coasts, attacking peasants and stealing their food. By the end of 2029, there was no functioning central government in Mexico, and in large parts of the country, no rule of law either. General Geronimo Morelos Paras reasserted control over central and southern Mexico in the late 2030s (most of the north and northwest remained under the control of local governors-turned-warlords), but by 2040 the population in the former national territory had fallen from 2029's refugee-swollen total of 155 million to only 115 million. And most of them were still hungry.

None of the Caribbean island states, except Haiti, suffered comparable population losses in the first half of the twenty-first century, partly because the drying effects of global warming were less pronounced on subtropical islands surrounded by oceans, and partly because there were still some fish in their oceans. The consequences in Cuba were the most surprising, however, for nobody had expected the communists ever to make a comeback there, despite the brutal impact of the post-Castro "democratic" regime on the living standards and life prospects of the poor.

The unforeseen effect of the creation of a virtual but impermeable border between Cuba and the United States was to loosen the grip of the Miami Cuban elite on the island. As U.S. citizens, they were still free to come and go, but the psychological impact of that heavily armed frontier on both the Florida Cubans and the real Cubans was very great. The great land scams of the years immediately following Raul Castro's death in 2016, when huge tracts of state-owned land were bought up by wealthy Cuban-Americans while local campesinos ended up landless and even poorer than before, had left a deep well of resentment.

The closure of the U.S. border made the incomers seem somehow more vulnerable than before, and there were still plenty of communists around. Indeed, some of them still had their old militia weapons, and now, for the first time in a very long time, they actually had a measure of popular support. The communist coup of August 16, 2028, was greeted by an explosion of joy in the poorer parts of cities and towns, from one end of the island to the other—and the U.S., with a great deal of other Latino business on its plate, simply accepted it. American troops were withdrawn from all parts of the island, except Guantanamo Bay, within a month.

The problem that distracted the attention of the U.S. government was this. Back in the year 2010, the Mexican-American population had amounted to about thirty million people, a large majority of whom were American citizens. Adding about five million other people whose families came from Spanish-speaking Caribbean or Central American countries, the total share of the U.S. population who were of Hispanic origin was then about 12 percent, or one in eight. Moreover, most of them had preserved their cultural ties and their language into the second and third generations much better than typical immigrant groups.

They were already a very powerful political lobby, especially in the border states where they were most heavily concentrated, and their influence had been almost as important as that of the agribusiness interests to keep the U.S.-Mexican border porous. It was a striking phenomenon of the early twenty-first century that not only Mexican-American groups, but the Mexican government itself, would make outraged protests if there were any official U.S. attempt to clamp down on illegal border crossings, as if undocumented Mexicans had a natural right to enter the United States unobstructed. But in the next two decades, as southern Mexico and Central America heated up and dried out, the number of illegal crossings soared. By the time that the U.S. government responded to popular panic and started building the Big Fence in 2027, the resident Mexican-American population, including recently arrived illegals, amounted to about 60 million people out of a total of 360 million: one in six. Closing the border was bound to cause a fight, but nobody realized how big it would be.

> Political and security systems in the world actually have environmental niches of their own. Our environment has changed so slowly that we don't really think of this, but today's systems, no matter how different they are from one another, have all evolved in the same climatic era, they all fit into circumstances that ultimately are governed by · the kinds of climates in which human cultures can exist, and it follows that if there are changes in the climatic background that are sufficiently vigorous, something is going to happen to the political systems as well.
>
> These kinds of stress[es] in one place or another, in one form or another, do not mean good news for representative democracy.
>
> —Leon Fuerth, professor of international affairs, George Washington University, in an interview with the author, February 5, 2008

At first the Mexican-American community (and the other, much smaller Hispanic groups) split down the middle, with the longer-established families tending to agree with the non-Hispanic majority that something drastic had to be done about the border. But when they saw how drastic it was going to be — *had* to be, if it was going to work — they were appalled. The pictures of innocent people being blown up by land mines and machine-gunned by automated weapons had the same impact on Mexican-Americans that it had on Mexicans south of the border: shock, horror, and an outraged rejection of a policy that involved such cruelty. Never mind that the families had voluntarily put themselves in harm's way, or that the United States had no obligation to take them in. Emotionally, it was just unbearable to watch, and it drove Mexican-Americans into an almost united front of impassioned opposition to the whole policy.

For a great many other Americans, unfortunately, this led directly to the conclusion that Mexican-Americans were a fifth column in the service of a foreign government. If the American people had been feeling more secure and more confident about the future, they might have managed a less hysterical response to the situation, but they had been through a lot lately: the permanent loss of some of the southeast Florida coast after Hurricane Tyrone; the chronic water crisis that hobbled all of the big southwestern cities, and some of the southeastern ones as well; the storm surge that had ruined most of the low-lying farmland around Chesapeake Bay and temporarily flooded some of downtown Washington; even the big San Francisco earthquake, though that had nothing to do with climate change.

There were plenty of voices of reason urging calm and compromise. One of the most prominent was that of former president George W. Bush, now eighty-four years old and very frail, who emerged from his long seclusion to set up a months-long video blog pleading with people to understand them. Even people who remembered his foreign policy fiascos of a quarter-century before and still reviled him for them were moved by the old man's persistence, but it didn't actually do any good. Americans were just not in an understanding mood, and they were particularly not in the mood to understand Mexican-Americans.

It never grew into what you would call a civil war, although lots of people died in the riots and in clashes between smaller groups and individuals, and a lot of people moved out of "mixed" neighbourhoods in order to feel safer. In almost all the counties along the border itself, Mexican-Americans were in the majority, and inevitably some young hotheads would go out to take a potshot at the crews building the Big Fence from time to time. Since the crews were also being shot at from the Mexican side by individuals wishing to demonstrate their disapproval of the whole concept, their responses tended to be heavy-handed, and kids were killed. After two or three years, with the border closed and Mexico in meltdown, the angry rhetoric in the United States died down and a sullen version of normal service was restored, but something important had been lost. For the first time since the Civil War, a major American ethnic group other than African-Americans felt permanently alienated from the mainstream culture and its assumptions and beliefs.

Mexican-Americans and other Hispanic-Americans still had higher birth rates than most other American groups, and were likely to end up as about a quarter of the population by the end of the century, but they were becoming a more and more separate group, which did not augur well for the future. If this was the way Americans responded to the first big crisis of the new era, how would they cope when the full effect of climate change hit the country?

CHAPTER THREE
Feedbacks: How Much, How Fast?

THE FIRST TIME I HEARD SOMEBODY SAY "The carbon sinks are failing!" I thought of the scene in the 1973 eco-disaster movie, *Soylent Green,* set in a grotesquely over-crowded 2022, when almost all of nature has been swept aside in order to feed the masses, and in which the onset of terminal collapse is due to the fact (a closely guarded secret) that "the plankton beds are failing." Life has a bad habit of imitating art. But while the real world will not be as hot, nor food as scarce, in 2022, as the world of *Soylent Green,* the prospects for a decade or two later are a good deal worse than most people realize. The conclusions of James Hansen et al., discussed in the previous chapter (pp. 43–75), tell us the probable long-term implications of doubling the amount of carbon dioxide in the atmosphere, but not the speed at which the average global temperature will rise on the way to that ultimate, catastrophic destination. Unfortunately, this is the realm in which those damnable "non-linear processes" hold sway—which means, as the scientists keep telling us, that we should expect unpleasant surprises. They cannot tell us exactly when these will happen, that being the nature of surprises, but they can and do tell us what is most to be feared, and at approximately what stage in the warming process it is likeliest to occur. The most unwelcome of these "surprises" are the feedbacks that they think may kick in at an average global temperature somewhere between two and three degrees Celsius hotter than the pre-industrial level.

The current (2007) *Intergovernmental Panel on Climate Change Fourth Assessment Report* does acknowledge and incorporate a couple of the feedback mechanisms that are expected to occur as the temperature rises. One is an increased rate of evaporation from warmer oceans, which will lead to more warming, since water vapour traps heat: a positive feedback. ("Positive"does not mean "good" in this context; it means "more"—as in "more heat.") On the other hand, the same increased rate of evaporation ought to lead to more cloud cover over the oceans, which would reflect more sunshine away from the planet's surface: a negative feedback that should, in theory, tend to bring the Earth's average temperature back down. (The argument over which of these phenomena will outweigh the other is still not concluded, but recent field research suggests that the evaporation will cause less cloud cover than the IPCC hoped.)

Another feedback, this time entirely positive (that is, tending to speed the warming), is the melting of the polar ice cover, particularly on the Arctic Ocean and around the Antarctic coast. This melting near the two poles will replace highly reflective, white ice sheets, which bounce 70 percent of incoming sunlight straight back into space, with open ocean that absorbs 94 percent of the sunlight striking it and converts it into heat.

Both of these phenomena can be modelled to some extent already, so they do get included in the IPCC Report. Unfortunately, the really big and dangerous feedbacks are left out of the IPCC Report entirely, except for the note that explains that they were not included in the calculations because there is no way yet available to model these processes. Because it ignores these feedbacks, the best-guess predictions in the IPCC's current report are on the order of a couple of degrees Celsius hotter by 2100. Unfortunately, the key issue is precisely the feedbacks that have been omitted, and which are already starting to happen in small ways. They would gain unstoppable momentum when average global temperature reaches somewhere between two and three degrees Celsius hotter.

[These] feedbacks obviously include the release, due to warming, of vast amounts of fossil methane hydrate from the Arctic, and also the release of methane hydrate from the oceans. There are vast melt-lakes in northern Canada and in Siberia where the methane is bubbling out so fast that it's difficult for them [the lakes] to refreeze. This methane is twenty-two times worse than carbon dioxide [as a warming agent], and so warming will be much faster as you include this feedback mechanism. There is another feedback mechanism, which is the fact that this same tundra possesses vast amounts of fossil carbon dioxide, and so, as the tundra warms, it not only gives up the methane, it gives up the carbon dioxide.

There is another issue in terms of feedback, which has to do not with carbon dioxide sources but [with] carbon dioxide sinks. This has to do with the fact that about half of the carbon dioxide sink is in the oceans. This is the absorption of carbon dioxide by the water and its utilization by the algae. What's happened is that because we've planted winter wheat and changed the land cover, vast amounts of iron-rich dust no longer drift out over the ocean, and algae requires iron-rich dust. So, 30 to 40 percent of the algae [has] gone over the last several decades, say the NASA overheads, and also because of the carbon dioxide absorption into the water, the water is turning into carbonic acid, which is further killing the algae. The bottom line is that the estimates now, from recent measurements, are that about half of the carbon dioxide uptake in the oceans is gone.

So, there are several mechanisms all producing much faster warming, much more intense warming in terms of these feedback cycles, not included in the IPCC estimates. These effects then cause even further evaporation, and water vapour is a greenhouse gas, and also further changes in the ice. If you take all these feedbacks into account, the estimates appear to be that by 2100, instead of two degrees or so Celsius rise, with a maximum of six degrees, it looks like a possibility of six to twelve degrees. At the four- to six-degree-plus range, all the ice eventually melts, and if all the ice eventually melts, we're looking at a sea-level rise of, instead of one to two metres, about seventy to eighty metres. Plus, [these kinds] of temperature changes would change the ocean-circulation patterns and end up with much of the oceans going anoxic—very low oxygen content—which would then promote bacteria which produce hydrogen sulfates, and these would rise and take out the ozone layer, and also make it somewhat difficult to breathe. This is by 2100.

—Dennis M. Bushnell, chief scientist,
U.S. National Aeronautics and Space Administration,
in an interview with the author, February 2, 2008

In *The Revenge of Gaia*, the independent scientist James Lovelock recalls a meeting of climate scientists in 1987. With their limited knowledge of the climate system at that time, they plotted three possible paths for global warming: a lower scenario in which the world grew hotter at the very modest rate of 0.06 degrees Celsius per decade, a middle scenario in which it warmed at 0.3 degrees Celsius per decade, and a high scenario in which we experienced 0.6 degrees Celsius of heating per decade. The reality places us today closer to the high than the middle scenario on that graph—and if you project that line into the future, it yields a climate where we hit two degrees Celsius hotter by 2020. That is not at all incompatible with a forecast of six to twelve degrees hotter by the end of the century.

These are not numbers that any political leader in the world acknowledges. The two candidates for the U.S. presidency in 2008 both promised dramatic cuts in American greenhouse-gas emissions—50 percent for Senator McCain, 80 percent for Senator Obama—but only for the year 2050. Eighty percent cuts in emissions by 2050 is also the announced target of the British government, and the European Union has spoken in similar terms, as has that most unlikely of green heroes, Governor Arnold Schwarzenegger of California. If the point of no return, after which the world is irrevocably committed to large and potentially runaway temperature change, is likely on a "business as usual" basis to be passed by 2020 or 2025, then they should probably be talking in much more urgent terms. But they don't, in some cases because they don't know how fast things are now moving and, in other cases, because they are reluctant to question the accuracy of the only numbers on climate change that have now gained widespread public acceptance: the IPCC numbers.

In the past couple of years, there has been a huge shift in Western public opinion on the issue of global warming, with large majorities even in the United States now identifying it as a serious problem. The main point of reference for the discussion has been the reports issued by the Intergovernmental Panel on Climate Change every five or six years, and, in particular, its fourth report, issued in four parts over the course of 2007. Amidst the cacophony of the debate about the once-controversial phenomenon of anthropogenic (human-caused) climate change, the IPCC reports have attained this exalted status over the years precisely because they are very cautious and conservative.

We really should not be surprised at the widening gap between the IPCC report and reality, for the IPCC's estimates for how much global warming we should expect, and how soon, have fallen far short of the reality from the start. The IPCC is, as the title suggests, an inter-*governmental* organization involving all the major countries of the world. The executive summary (the only part of the report that lay people, including most journalists, will ever read) is edited by the participating governments in consultation with the scientists who are lead authors of the various parts of the report, but the governmental involvement starts well before that.

In the end, it comes down to what the scientists believe is solid evidence, and it's essentially based on a review of the peer-reviewed literature in reputable scientific journals. Where the governments come in is . . . first, in choosing some of the lead authors; secondly, in helping to define the scope of each report; thirdly, in providing review comments at different stages of the process, and finally, in approving the summary for policy-makers, which is a very tortured process, I would say, and one in which some degree of flexibility and compromise is required with the lead authors. But, in the end, the science is approved by the scientists, with wording that is suitable for governments.

> —Kevin Hennessey, principal research scientist,
> CSIRO Marine and Atmospheric Research,
> and coordinating lead author of the Australia-New
> Zealand chapter of the Working Group II report of the IPCC,
> in an interview with the author, May 12, 2007

It is a good thing to have an institution like the IPCC, which periodically brings together several thousand leading scientists from all over the world, working in the many disciplines relevant to the question of climate change, to undertake a methodical assessment of the evidence on the probable speed and scale of global warming. It is not even a bad idea to have the world's governments involved in the process, as that guarantees that they will take a close interest in the outcome and even, to some extent, feel bound by it (although the IPCC's remit, in Kevin Hennessey's words, is to be "policy-relevant but not policy-prescriptive").

However, governmental involvement does imply an inbuilt conservatism in the "policy-relevant" conclusions of the IPCC, for governments do not want reports that force them into major unanticipated expenditures. Another reason for the conservatism is the natural dynamics of a consensus-based process, in which the scientists on each committee are much likelier to reach consensus at the lowest common denominator of a range of estimates about the severity of the problem than at the high end of the range. A third reason is the fact that the IPCC (which does not do original research) imposes a cut-off date after which newly published scientific studies cannot be included in the material considered by the various working groups. Fair enough, as they'd never get the reports finished otherwise, but it does mean that the latest evidence is never included in the reports.

For the 2007 report, the latest publication date for scientific papers to be included among those considered in the IPCC process was the end of 2005. But the data being analyzed in those papers are generally a good deal older than that, for two reasons. One is the publication process itself. The first step in publishing a scientific paper is to submit it to a scientific journal, whose editor then sends it out to other scientists with established reputations in the field for peer review. The other scientists reviewing the paper will often raise questions that are then referred back to the authors for comment. There may be two or more iterations of this process before the editor of the journal, if he or she is satisfied, puts the article in question into the queue for publication—which is frequently a year or more long. Realistically, the great majority of the scientific papers that were published before the end of 2005 were probably submitted for publication before the end of 2003.

The other problem is that the data are by definition older than the paper. The data series that constitute the foundation of a given paper will normally end some number of months, or even years, before it is written and submitted for publication. It takes time to analyze the data and draw conclusions, and the best scientists are very busy people. The net result is that most of the data that went into the scientific papers that formed the basis for the IPCC's 2007 report actually refer to 2002 or earlier. In 2008, we are working with predictions that essentially take little notice of the evidence of the past six years.

There is no scientific failure in all this. It's hard to see how peer review could be speeded up drastically without damaging the integrity of the process, and a cut-off point is necessary because you cannot expect the IPCC's committees to be responding to a daily avalanche of new data while they wrestle with a mountain of information that has already accumulated since they last met. Leaving the biggest potential feedbacks — methane and carbon dioxide release from thawing permafrost in the higher latitudes, and carbon dioxide release from warming oceans — out of the climate change scenarios that the IPCC generates is defensible in scientific terms, for they did genuinely lack the ability to model them accurately. But, in a report intended for non-scientists, this omission ought to have been highlighted in warning yellow, not buried in the footnotes. The fact that this huge hole in the middle of the analysis was *not* flagged more prominently may be due to the influence of governments on the IPCC process, or it may just reflect the naïveté of scientists about how non-scientists treat evidence.

At any rate, the big feedbacks are the real danger, and a lot of recent evidence, none of it considered by the scientists who wrote the IPCC's 2007 report (though many of them knew about it) suggests that these processes are already beginning to be activated.

> The feedbacks in climate change can operate in both directions, but what we are seeing is more positive feedbacks than negative ones . . . A good example of those positive feedbacks is that, as the Earth gets warmer, the ocean is actually able to hold less carbon dioxide, so the ocean takes up less carbon dioxide out of the atmosphere . . . It's like the principle that warm beer goes flat. We're talking about a carbonated ocean, and it gets less carbonated as it gets warmer.
>
> We're seeing, similarly, a reduction in the rate at which terrestrial ecosystems, forests and soils, take carbon

out of the atmosphere . . . The amount of carbon dioxide that is being removed from the atmosphere, both by the oceans and by the forests and soils, is actually going down at the same time that the emissions are going up. What stays in the atmosphere is the difference between the emissions and what the oceans and the soils and the forests take out, so that increment, that addition, has been growing very rapidly. The emissions are going up, and the so-called sinks, the things that draw the carbon dioxide out of the atmosphere, are diminishing in their potency.

Methane is particularly problematic because it is about twenty times as potent per molecule as a greenhouse gas as carbon dioxide is, and there is a lot of methane trapped in compounds called methane clathrates, basically ice crystals with methane in the middle of them, under the permafrost as well as under shallow seas. The potential for a large positive feedback, by driving a lot of methane into the atmosphere as the Earth gets warmer, particularly now from the permafrost, from the frozen North, is very large and very troubling. There is evidence that the methane emissions from the far North are increasing. It's preliminary and spotty evidence, but it's all in the same direction. It's all in the direction of indicating that the theory that says this could be a problem is becoming an experimental reality.

—John Holdren, director, Wood's Hole Research Center,
and past president, American Association for the Advancement
of Science, in an interview with the author, February 6, 2008

In sum, nobody can tell you with full confidence just how much worse than the IPCC scenarios the real situation is, but serious scientific opinion thinks that it is quite a lot worse. There are, for example, an estimated seventy billion tonnes of frozen methane gas in the west Siberian bog alone—and the southern fringes of it began to melt in 2005. We have less time to respond than we thought, and the emission cuts we must make will need to be much deeper than the conventional wisdom decrees if we are to avoid runaway change. For that is the ultimate danger: that somewhere not too far beyond two degrees Celsius warmer than the pre-industrial global average temperature, the feedbacks will create a self-sustaining process of further warming that we cannot stop, no matter how deeply we cut our emissions subsequently.

The European Union has officially adopted two degrees Celsius as the ceiling beyond which global warming must never go, and many, probably most, climate scientists in other parts of the world would accept that as the best available target. The number is not chosen arbitrarily, but there are elements of guesswork in it, as well as a pragmatic recognition that we are extremely unlikely to stop the warming process short of there. In terms of the concentration of greenhouse gases in the atmosphere (which is the measurable target that we must actually aim at), a maximum of two degrees Celsius hotter than the pre-industrial average global temperature is generally believed to translate into 450 parts per million of carbon dioxide equivalent in the atmosphere. As we have seen in the previous chapter, this should probably be regarded only as a maximum from which we must then retreat as rapidly as possible, heading back down towards 350 parts per million or even lower, if we truly want to stay in the Holocene, but let us accept it as this generation's goal.

We should also consider two clarifications and a significant fact. "Carbon dioxide equivalent" refers to the concentration of *all* greenhouse gases in the atmosphere, the others gases (such as methane and nitrous oxide) being expressed in terms of how much carbon dioxide would produce the same warming effect. The "measurable target" of 450 parts per million carbon dioxide equivalent in the atmosphere is *not* a bankable guarantee that we will be all right if we manage to slide in under it at the end of this century: the science is not that precise.

It is just a target that offers us a 50 percent probability of stopping global warming at two degrees Celsius. But the atmosphere already holds 387 parts per million of carbon dioxide now (in 2008), not to mention the other greenhouse gases that contribute to the total "carbon dioxide equivalent." And the current annual rate of increase in atmospheric carbon dioxide (which is accelerating as emissions rise rapidly in the developing countries) is nearing three parts per million.

Clearly, we have very little margin left to play with, and if global emissions are not radically reduced within about twenty years, then we will have little chance of stopping short of 550 parts per million carbon dioxide equivalent. Fifty or sixty or even eighty percent cuts in emissions by 2050, which is the target that has won the hearts of most contemporary politicians who take climate change seriously, gets us to 550 parts per million at about the same time. Average global temperature might still only rise to two degrees Celsius warmer by 2030, or even less if the feedbacks are slow to kick in, because there is a very long lag between the arrival of a given amount of carbon dioxide in the atmosphere and the consequent change in the planet's average surface temperature. But once the carbon dioxide is in the air, it stays there on average for about two centuries, so the final outcome has become almost inevitable. A concentration of 550 parts per million carbon dioxide equivalent in the atmosphere takes us to a world that is between three and four degrees Celsius hotter by the end of the century, *not counting* any feedback effects. But, by three degrees Celsius hotter, the major feedbacks surely will have begun to operate, and our goose will be well and truly cooked.

There is another, more historical way of looking at all this, though it leads ultimately to the same conclusion. The pre-industrial concentration of carbon dioxide in the atmosphere circa 1800 was about 280 parts per million. As the industrializing nations of Europe burned more and more coal in the nineteenth century, the concentration began to rise, but only very slowly, for their total population at the time was under three hundred million. In the late nineteenth century, North America and Japan also industrialized and, as the twentieth century dawned, oil and natural gas were added to the fossil fuel mix, but even as late as 1960, all of their activities were adding less than one part per million of carbon dioxide equivalent to the atmosphere each year. At that time, mass car ownership was still exclusively a North American phenomenon, and only one-quarter of the world's population lived in countries that could be described as "developed."

Between 1960 and now, the world's population has more than doubled, and a growing share of that larger population has entered the phase of rapid industrialization. About four billion of the world's current total of 6.7 billion people live in countries that are either already fully industrialized, or that are experiencing economic growth rates that consistently exceed 5 percent a year, which means that they should arrive in the industrial promised land (other things being equal) in one or, at the most, two generations. The total population of all the countries that emitted large amounts of carbon dioxide per capita in 1960 was well under one billion. Now it is four billion, so global greenhouse-gas emissions have grown accordingly. The rapid rise in carbon dioxide emissions is all post-1960, but it has been extremely rapid.

As a result, we have experienced a rise in global average temperature on the order of 0.8 degrees Celsius since the start of the industrial revolution around 1800, almost all of it occurring within the past thirty years. Moreover, the greenhouse-gas emissions that have been put into the atmosphere in recent decades already commit us to a further warming of 0.4 or 0.5 degrees Celsius over the next twenty years or so. That makes a total of 1.2 or 1.3 degrees of warming already. If we want to have any hope of sliding in under two degrees Celsius hotter at the point where we finally get our emissions under control, we have just over half a degree left to play with. All of our emissions in the rest of this century—and indeed, in the rest of human history—must not drive the atmospheric concentration of carbon dioxide above this level.

> JAY GULLEDGE: The point is that there's this large store of a powerful greenhouse gas, literally many, many millions of tonnes of the stuff, that is close to the surface of the Earth. It's frozen in place right now, [but] we have a warming trend, and if you just put two and two together, this means that there's some risk of releasing that methane, and if it were to happen very quickly, it would produce a very strong kick to the warming that's happening now.
>
> If that happens, we will have lost control over the greenhouse-gas content of the atmosphere. Right now, we are driving that. Our activities are dictating that. If we reach a point where the positive feedbacks take over, we no longer have the option of intervening, changing our activities, and then ramping down the greenhouse-gas content of the atmosphere.
>
> GD: What's the point of no return here?
>
> JAY GULLEDGE: I honestly think we don't know that, but I think we can give a range within which we get into a risk of hitting a point of no return, and then as you go farther into that range the risk grows . . . Somewhere between one and four degrees above where we are today—somewhere in that range we'd probably cross a threshold at which we wouldn't be able to stop the disintegration of Greenland and West Antarctica [ice sheets] . . . One to four degrees means that we would probably hit that threshold sometime in the next forty, fifty years. After that, all bets are off as to whether we still have a chance to pull it back.
>
> —Jay Gulledge, senior scientist and program manager
> for science and impacts, Pew Center on Global Climate
> Change, in an interview with the author, February 4, 2008

This chapter is really about the timetables of the crisis. When do the feedbacks hit? When does food production start to plummet? When do the political reactions become so extreme that global solutions are no longer possible? Of course, the timetables cannot be established with any confidence: there are too many variables, both physical and human, to allow any precision in the forecasting. Nevertheless, a cautiously pessimistic estimate, allowing for recent evidence of the speed with which feedbacks may be coming into play, still suggests that climate-related disasters large enough to make further global cooperation unlikely will probably not occur within the next fifteen to twenty years. (We might well slide past the point of no return for runaway climate change during that period if the measures we take to curb our emissions are not radical enough, but we would not actually experience the resulting disasters until a good deal later.) There are, however, two commodity-related crises that could strike a lot earlier, and severely narrow the scope for global cooperation. One has to do with oil, the other with food.

In March 2008, oil breached the US$100 per barrel mark for the first time. At the time of writing (July 2008) it was nearing US$150. There is not yet an absolute scarcity of oil, in the sense that supply cannot physically match demand at any price. Indeed, we are probably still a long way away from that situation, for in all previous crunches of this sort, higher prices eventually reduced demand, and so brought supply and demand back into equilibrium (although the poorest consumers of oil, at that point, were simply out of the market). But two things are different in the current situation.

One is that higher oil prices are *not* likely to stimulate enough new exploration in difficult and expensive environments to increase supply substantially, as they did after a five- to ten-year lag in previous episodes of this sort. The consensus of expert opinion in the oil business is that the last major conventional oil fields were probably discovered in the 1970s and

1980s, with the possible exception of resources in the Arctic Basin that will not become accessible until more of the ice melts. Smaller, more difficult fields in deep water or remote locations will continue to be found; improved recovery techniques will continue to squeeze more oil out of old fields; and "unconventional oil" from tar sands, oil shales and the like will continue to be developed. It is not the end of oil; as an insider in the Calgary oil patch put it, it is the end of "cheap, sweet oil." But none of this will fully compensate for the gradual decline of the mega-fields that fuelled the great twentieth-century blowout. Oil prices will continue to fluctuate, but we have already seen the end of cheap oil.

The other difference in the current situation, closely related to the first, is the widespread conviction that somewhere, not too far ahead of us, lies the moment of "peak oil." The theoretical underpinning for this is the famous "Hubbert curve," named after the American petroleum geologist M. King Hubbert, who observed that production from any given oil field tended to peak about thirty-five years after it was discovered. Extrapolating this observation to the United States as a whole, where the discovery of new oil reserves peaked in the 1930s, he predicted in the 1950s, in the face of almost universal disbelief, that domestic American oil production would peak between 1965 and 1970 (it was actually 1970), and subsequently decline. Extrapolating again to global oil production, the Hubbert curve suggests that since the discovery of new oil reserves peaked worldwide sometime in the 1970s or early 1980s—there is some room for quibbling on the precise date, since the oil price crises of 1973–74 and 1979–80 stimulated one last big boom of exploration and discovery—the actual peak of world production should occur around thirty-five years later. That is to say, around now, or not very far from now.

This should not imply instant disaster. The typical graph of production over the lifetime of a large oil field is well established, and it is quite symmetrical: production rises rapidly at first, then more gradually. In the years before it peaks, it is rarely rising at more than 2 percent a year. The fall in production, after the peak has passed, is usually just as gentle: rarely more than 2 percent a year. Modern recovery measures may mean that the fall in production in some places is significantly delayed at first, and then far steeper at the end, but it is probably still safe to assume that peak oil would be followed globally not by a crash in oil production, but by only a gradual, prolonged decline. The problem is that world demand for oil continues to rise at around 2 percent per year, driven mainly by the rapidly growing economies of East and South Asia. A 2 percent annual fall in production, plus a 2 percent rise in demand, adds up to a cumulative annual 4 percent shortfall in supply, and that is serious. In fact, it is potentially lethal.

The problem is that nobody believes in the market. Theoretically, the international market for oil, with prices set moment by moment in London and New York for the various grades of crude, should efficiently allocate oil to those customers best able to pay for it. The great powers, almost by definition, would be the main beneficiaries of this process, since you don't get to be a great power without having very deep pockets. However, the great powers are also precisely the countries that have the bargaining strength (and the military strength) to subvert the market—and they succumb to the temptation to do so all the time. These are the customers with the purchasing power to draw the producers into long-term, fixed-price contracts that avoid the vagaries of the market and ensure a steady supply of oil, regardless of those fluctuations—and they are the countries that can (and sometimes do) use their military strength to freeze rivals out and ensure that the oil comes to them.

The major Western powers have played this game (mostly against one another) for about a century already, and now new players, such as China and India, both with high and growing demand for oil imports, are joining the competition. For example, the Chinese regime concluded, rightly or wrongly, that the

American invasion of Iraq in 2003 was an attempt to create a permanent American military bastion in the Gulf, the source of almost all of China's imported oil, which would enable Washington to cut off the flow of oil to China in the event of a serious confrontation between the two countries. China's response has been to try to diversify its sources of oil by creating allies and client states among the oil-producing countries of Africa. (One tragic consequence of that has been the Sudanese regime's ability to defy Western and African pressure to end the killing in Darfur.)

The sane way to respond to a post-peak oil situation, in which supply was falling short of demand at the incremental rate of 4 percent a year, would be to let the market set the price; give poor countries emergency financial aid to enable them to import essential supplies; and work very hard to reduce the domestic demand for oil in the developed countries by more than 4 percent a year. Achieving the latter goal would probably be possible if enough money were thrown at the problem — and once demand was falling faster than supply, the price of oil would tend to come back down. But history suggests that the collective response of the great powers to peak oil may be to plunge into a race for control of the major remaining sources of oil beyond their borders, with much use of military force in the major oil-producing regions, and an overarching strategic confrontation that broadly resembles the old Cold War. In such an international environment, there would be little chance that any of the deals and treaties that are needed to cope with the threat of global warming could even be negotiated, let alone enforced. Indeed, we are already beginning to see the faint outlines of a similar pattern in the early international response to the other looming crisis of scarcity: that of food.

> Abu Dhabi is developing nearly 30,000 hectares of farmland in Sudan in the first step towards ensuring food security in the emirate. The move follows similar projects by

Middle Eastern countries locking up land from Brazil to Pakistan and Thailand to guarantee supplies of cereals, meat and vegetables at a reasonable cost. Mohammed al-Suwaidi, the director of the Abu Dhabi Fund for Development, which runs the farms, said yesterday that food security was a priority for his government, and that the Sudan deal "will not be the last project. . . . The recent oil price boom had a major effect on the price of raw commodities," Suwaidi told Reuters. "Global warming has an effect on commodities. The time may come when, even if you have the money, acquiring some commodities will not be easy."

Crops will include corn, alfalfa, and possibly wheat, potatoes and beans. Everything will be exported to the UAE. Projects in Uzbekistan and Senegal are also being considered. China has been leasing farmland in Asia and Africa for more than a decade. Egypt and Saudi Arabia have also expressed interest in developing projects in Sudan, where production has been hindered by decades of conflict and misrule. . . .

Saudi Arabia is especially keen to secure its food supply, having recently announced the phasing out of a 30-year project to grow wheat in the desert because the water consumption was too high. Like Bahrain, it has announced plans to stockpile basic foodstuffs, and intends to invest in agricultural projects and companies in Brazil, Ukraine, Thailand and India . . . but questions remain about how the deals will work in the event of domestic shortages in the food-producing countries. . . .

Guardian report by Xan Rice, 23 June 2008

For the fifty years between 1950 and 2000, as the world's population more than doubled, grain production more or less kept pace—but then it stalled. In six of the past seven years, the human race has consumed more grain than it grew. World grain reserves at the start of the Northern Hemisphere harvest in 2007 were only 57 days, down from 180 days a decade ago.

To make matters worse, the demand for food is growing faster than the population. As incomes rise in China, India and other countries with fast-growing economies, consumers include more and more meat in their diet. The average Chinese citizen now eats fifty kilograms of meat a year, up from twenty kilograms in the mid-1980s—and producing meat consumes enormous quantities of grain. It takes seven or eight kilograms of grain to produce one kilogram of beef, and even one kilogram of chicken requires three kilograms of grain.

Of course, there is a not-very-secret escape hatch here. All the calculations of imminent food scarcity at the global level depend on the assumption that the meat-rich North American and European diet remains unchanged, and that the newly enthusiastic meat eaters of China, India and elsewhere have free rein to shift the mix of their diets in the same direction. Should they all suddenly get religion and turn vegetarian, the equations would change radically—and it wouldn't even harm the health of the neo-vegetarians. Unlike the involuntary vegetarians who populated the lower ranks of the early mass civilizations, and paid for their stunted diet with deficiency diseases and an average stature that was around 15 percent shorter than that of their meat-eating hunter-and-gatherer ancestors, modern vegetarians can maintain healthy diets.

> TIM LANG: As China and India are getting richer, they're eating more of a Western diet, so their footprint is growing exponentially. It doesn't add up. It can't go on in the way that Britain eats or America eats. Britain's food system is a six-planet food system. Britain consumes food as though it's got six times its land mass. It can't do it. Well, it can, but only by using other people's land [to produce a lot of the food on]. Why shouldn't they have it?

GD: So can we . . . continue to accomplish the one miracle we have accomplished, which is to feed three times
as many people off the same land?

TIM LANG: The calculations—back-of-an-envelope,
mostly—that have been done so far show that if you alter
the diet, the capacity to feed the population rockets up.
You can't feed the world's population if you eat like the
British. Not a chance if you eat like the North Americans.
Not a chance. But if you eat much less meat, much less
dairy, suddenly the possibilities grow. The variations in
how much land you have available and how many people
you could feed all depend upon what diet they eat.

—Tim Lang, professor of food policy, City University,
London, in an interview with the author, March 25, 2008

So we wouldn't be facing a world food-supply problem soon
if we all become vegetarians tomorrow—but we aren't going to,
are we? Besides, meat-eaters are not the primary cause of the steep
rise in grain prices in the past few years. Two other factors—global
warming and biofuels—are responsible for that, and at this point
in the game global warming is almost certainly the lesser of them.

Global warming is probably already cutting into food production in some areas. For example, many people in Australia,
formerly the world's second-largest wheat exporter, suspect that
climate change is the real reason for the prolonged drought that
is destroying the country's ability to export food. But we would
still have some room for manoeuvre, if it were not for the rage
for biofuels. Thirty percent of this year's U.S. grain harvest will
go straight to ethanol distilleries, and the European Union is
aiming to provide 10 percent of the fuel used for transport from
biofuels by 2010. A huge amount of the world's farmland is
being diverted to feed cars, not people.

If . . . more and more land [is] diverted for industrial biofuels to keep cars running, we have two years before a

food catastrophe breaks out worldwide. It'll be twenty years before climate catastrophe breaks out, but the false solutions to climate change are creating catastrophes that will be much more rapid than the climate change itself.

—Vandana Shiva, director, Research Foundation for
Science, Technology and Natural Resource Policy, New Delhi,
in an interview with the author, March 18, 2008

The world food supply is heading into a perfect storm, and the key element that is pushing the system into crisis is biofuels. Or to be more precise, it is the first-generation biofuels, based on converting corn (maize) or sugar cane into ethanol, and soy or palm oil into biodiesel, that are now being produced on a large scale in the United States, Brazil, Europe, India and a number of Southeast Asian countries.

There is nothing wrong with the concept of biofuels in principle. There is an urgent need to replace fossil oil, which returns long-stored carbon dioxide to the atmosphere when it is burned, with non-fossil fuels of similar properties, suitable for powering vehicles of every kind, that do not add to the long-term burden of carbon dioxide in the atmosphere. Biofuels that absorb carbon dioxide as they grow, and release it again when they are burned, offer an attractive solution—*provided that* they do not displace food crops or forests, and that they really are carbon-neutral or close to it. Unfortunately, the present generation of biofuels meets neither requirement.

It's hardly surprising that most currently available biofuels are not carbon-neutral, since they were never intended to be a solution to climate change. When Brazil began converting sugar cane into ethanol in the 1980s, and when the United States followed suit by beginning to subsidize the conversion of corn into ethanol in the early years of this century, the stated purpose was, in President George W. Bush's phrase, "energy independence." (He didn't yet believe in climate change.) The notion was to replace imported oil, which was subject to sudden price fluctuations and to politically motivated interruptions of supply, with a home-grown source of fuel that was always available and not too expensive. At the time, nobody even calculated whether it was carbon-neutral or not. That (specious) claim was only made years later, when climate change had become a politically sensitive issue.

For most of the current generation of biofuels, however, the claim truly is nonsense. Even when the biofuel crops are being grown on existing farmland, the energy inputs involved in planting, tending and harvesting the crop and converting it into biofuel often mean that there are large net emissions of greenhouse gases. (Corn-based biofuel does particularly badly in this regard; sugar cane does better.) But when forest is being cleared to grow biofuels, as is the case in Amazonia and in the tropical forests of Asia, the equations involved become quite insane. A recent study in the U.S. journal *Science* calculated that destroying natural ecosystems to grow crops for biofuels releases between 17 and 420 times more carbon dioxide than will be saved annually by burning the biofuel grown on that land instead of fossil fuel. It's all justified in the name of fighting climate change, but the numbers just don't add up.

The second-generation or "cellulosic" biofuels, made from switch-grass, willow or other fast-growing plants raised on land marginal for agriculture, hold considerable promise for the mid-term future, and the third-generation biofuels, such as algae or "halophyte" plants that can be grown in stagnant or salty water, may be even better in the long run, but the current generation of biofuels is an almost unmitigated disaster. If the subsidies are not cut back and the farmland restored to food production, there almost certainly will be an absolute shortage of food in the world in only a very few years. In that case, the poor will be starving so that the rich can drive their automobiles on what they imagine is a more eco-friendly fuel. One cannot imagine a political environment less conducive to global cooperation on climate-change issues.

To the two commodity-related potential crises, oil and food, we can add a third that is more purely political: the risk that the United States and China might slide into a new Cold War that forces most other great powers to line up on one side or the other. This is not an imminent danger, and many people in both countries are working very hard to prevent it. Nevertheless, history suggests that periods when a long-reigning paramount power (like the United States) is being overtaken by a faster-growing rival (like China), they end up in military confrontation more often than not.

> The Chinese navy is poised to push out into the Pacific — and when it does, it will quickly encounter a U.S. Navy and Air Force unwilling to budge from the coastal shelf of the Asian mainland. It's not hard to imagine the result: a replay of the decades-long Cold War, with a center of gravity not in the heart of Europe, but, rather, among Pacific atolls that were last in the news when Marines stormed them in World War II.
> — Robert D. Kaplan, "How We Would Fight China: The Next Cold War," *Atlantic Monthly*, May 2005

There is a flourishing intellectual industry in the United States promoting the idea that a long military confrontation with China is inevitable. It has considerable tacit support from those branches of the U.S. Armed Forces that can only justify their large investments in high-tech military hardware by the existence of a "peer competitor": some other country big and powerful enough to justify the maintenance of twelve aircraft-carrier task forces by the U.S. Navy, for example. The only conceivable candidate for this role is China — and it is not to be doubted that there are similar groups in the Chinese armed forces who use the "American threat" to justify their own requests for more and better weapons. None of this necessarily leads to war, but the risk of a long military standoff is not negligible—and, if it occurred, then once again the probability of the kind of coordinated global action that would avert runaway climate change shrinks drastically.

It is absolutely clear that global cooperation on dealing with this most global of problems will only be possible in a relatively peaceful and non-confrontational environment. We have enjoyed such an international environment for a full twenty years now, since the end of the old Cold War: no great power seriously fears an attack by any other. These are not just the ideal conditions in which to try to negotiate a global deal on containing climate change; they are the indispensable conditions. If the great powers get into a desperate race to nail down dwindling oil supplies, if famine drives a wedge between those countries that can feed their people and those that cannot, or if the U.S. and China tumble into a new Cold War, there will be no global deal.

We really need this relatively fortunate time, when only minor issues like terrorism disturb international peace and order, to continue for at least another twenty years while we all work to get global warming under control. Whether we will actually be granted the extra time, unfortunately, is impossible to say. There are many unknowns in the field of climate change, though the sheer weight of scientific effort now being brought to bear on them is rapidly shrinking the areas of uncertainty. ("Twenty years ago we were in the Stone Age of climate change research," as Hans-Joachim Schellnhuber, director of the Potsdam Institute for Climate Impact Research, put it.) But the political, economic and strategic variables are even harder to calculate, and it is they that will decide whether human beings manage to contain the problem. The proposed remedies are numerous, but they don't all match up, and they almost all require that scarcest of commodities—political will.

NORTHERN INDIA, 2036

In December 2001, when terrorists attacked the Indian parliament, India blamed Pakistan and withdrew her High Commissioner in protest . . . At a seminar in Karachi in the last week of December 2001, attended by ICPI [International Center for Peace Initiatives], the only occasion when tensions rose was when someone alleged that the Indian government had plans to use the water weapon. A participant warned that any conflict over water would lead to Pakistan using nuclear weapons on a first strike basis against India.

— The Final Settlement: Restructuring India-Pakistan Relations, published by the Strategic Foresight Group, ICPI, Mumbai, 2005

THE SURVIVING MONKEYS STILL PLAY amid the ruins of the Taj Mahal. They come out of habit, although there are no longer any tourists leaving food around. Even if there were much left worth seeing on the site, the radiation levels are still too high. There are fewer monkeys than there used to be, too, for the blast was less than five kilometres away. It was a ground burst, to crush the aircraft in their shelters, and although it happened on the first day of the war, when both sides were still avoiding strikes against cities, it happened to Agra Air Force Station, located right at the western edge of the city. Maybe the Pakistani planner who put it on the target list didn't know about the Taj, or maybe he felt that he couldn't afford to care.

Flat and low-built, the city of Agra gave the Taj Mahal little protection from the blast. The three million people who used to live in the city had even less protection, and those who survived have mostly fled south, for the whole of northern India as far east as the city of Kanpur is still fairly radioactive, even three months after the short-lived nuclear exchange ended. In the final spasm of city-killing most of the big Indian cities were hit even in the south, but here in the north, every air-force base, every army canton-ment, everything of military value was targeted from the start. Foreign observers doing counts from satellites estimate that the corridor along the Grand Trunk Road from Allahabad up through Kanpur and Delhi to Amritsar, taking in most of the densely pop-ulated upper Ganges Valley and the eastern Punjab, was hit by over three hundred warheads in six days. That's still only one nuclear weapon for every five hundred square kilometres, on aver-age, but the fallout plumes overlapped almost everywhere.

Pakistan, from the satellite and UAV (unmanned aerial vehicle) reports, is even more grievously wounded, for it had less hinterland to start with. From Peshawar through Lahore and on down to Multan, the density of nuclear strikes was higher than in northwestern India. Further south in the Indus valley there were fewer strikes, but the cities of Sukkur, Hyderabad and Karachi all took multiple hits. You would think that there would be no functioning state on either side after all this—but you would, of course, be wrong. Like cockroaches, governments are largely immune to nuclear war.

New Delhi is a series of shallow, overlapping craters, but a provisional Indian government and a military command centre are attempting to reassert central control over the country from somewhere near Nagpur. The surviving Pakistani authorities are operating from bunkers near the ruins of the city of Rawalpindi. And while the ceasefire is holding, each still insists that the war was the other side's fault. Both are probably right.

Neither side would have behaved so obstinately if it were still able to feed its people, but the increasingly frequent failure of the summer monsoon was hitting Indian agriculture very hard, and both countries had been suffering for many years from the severe summer flooding of the glacier-fed river systems that rose in the Himalayas (the Indus, Ganges and Brahmaputra) as the glaciers melted. It didn't help, of course, that they had so many people who expected to eat well: there were well over two billion people in the Indian subcontinent by 2036, and a large proportion of them aspired to consume at a fully industrialized level.

Neither Pakistan nor India could produce enough food to cope with demand, and neither could buy enough on the international grain market to fill the gap, even with their newfound wealth. The international market had shrivelled to almost nothing as global warming hit agriculture right across the planet. Still, they staggered along somehow—until the summer flooding in the glacier-fed rivers turned to summer droughts.

It was perfectly predictable, and had been widely predicted for decades: first the glaciers will melt, overfilling the rivers every summer—and then they will be gone, and the rivers will run dry in the summers. It was a gradual process, and it didn't happen on every river at the same time, but by the mid-2030s the flow on most major rivers rising in the Himalayas was sharply down. This was a medium-sized problem for India, where a very large proportion of the crops is rain-fed. But it was a life-and-death crisis for Pakistan, a country that is essentially a desert with a big river flowing through it. At least three-quarters of Pakistan's food was grown on land that was irrigated by the Indus river system. It was, indeed, the largest single area of irrigated land in the world. And, by 2036, the Indus system was running on empty.

What turned an ecological problem into an international crisis was the fact that five of the six rivers that eventually feed into the Indus system rise in Indian-controlled territory. In undivided, British-ruled India, the water flowed unhindered into the intricately linked irrigation canals that covered much of the provinces of Punjab and Sind, but Partition in 1947 left most of the headwaters in Indian hands, while well over four-fifths of the farmers who depended on the water lived in the new state of Pakistan. Moreover, several of the rivers had their headwaters in the state of Kashmir, itself a disputed territory and the scene of the first Indo-Pakistani War. Fortunately, there was more than enough water available to meet everybody's needs: the mean annual flow of the Indus system in the first half of the twentieth century was 216 billion cubic metres, and, on the eve of partition in 1947, 98.7 billion cubic metres was still emptying into the Arabian Sea unused. India briefly interrupted the flow of water from its side in April 1948 in an attempt to force a new agreement, but there was no open conflict over the division of the waters and, after a dozen years of on-off negotiations, the two countries signed the Indus Waters Treaty in 1960.

It was a rough-and-ready division of the resource, mediated by the World Bank, that gave the three eastern rivers (Sutlej, Ravi and Beas) to India, and the three western rivers (Indus, Jhelum and Chenab) to Pakistan. This gave Pakistan over four-fifths of the total flow since the western rivers were much bigger (out of 216 billion cubic metres of water in the system, the three eastern rivers contained only 40.7 billion), but many Pakistanis still felt cheated: before Partition only about a quarter of the flow in the three eastern rivers had been used for irrigation on what was now the Indian side, and now India got it all. Moreover, India got the right to take enough water from the Jhelum and Chenab rivers to irrigate some 200,000 hectares of land. But Pakistan received substantial payments from India, supplemented by a large amount of foreign aid, to build new canals to redistribute water more efficiently among the western rivers, so everybody lived happily ever after—for a while.

But populations grew, especially on the Pakistani side, where the 34 million people of 1951 had multiplied to 170 million by 2008. By then, *all* the water in the Indus system was being used: in some months of the year the river did not reach the sea at all, and sea water was penetrating up to eighty kilometres into the estuary, causing increased salination in almost half a million hectares of farmland. The per capita availability of water to Pakistan's population had fallen from 5,300 cubic metres a year at Partition to only 1,000 cubic metres annually by 2008, a level that the United Nations defines as "critical," but at that point Pakistan could still feed itself.

Over the next three decades, Pakistan underwent a remarkably steep fall in its birth rate, reaching replacement level by the early 2030s. However, the very young age structure of the population, a legacy from the high birth rates of previous generations, meant that Pakistan's population still reached 290 million by 2036. If the country had continued to receive as much water as it did in the first decade of the twenty-first century, per capita availability of water would nevertheless have fallen to only about six hundred cubic metres a year — but, by the century's fourth decade, Pakistan was getting far less water from its rivers than before. The glaciers had almost all melted away, the water that used to fall as snow in the winter to replenish those glaciers now fell as rain and ran off immediately, and in the summers the rivers were only a shadow of their former selves.

Successive Pakistani governments did all the right things to alleviate the problem: barrages and dams were built to retain winter runoff for the summer, the irrigation canals were lined to reduce water loss through seepage, and drip irrigation systems were widely installed for the final delivery of the water. But nothing could really compensate for the fact that there was much less water in the rivers, and the worst hit was the Indus River itself. The Indus alone accounted for more than half the total flow of all six rivers in the system, but it was

90 percent glacier fed, and as the massive glacier on Mount Kailash in western Tibet melted away, its flow first soared and then plummeted. By 2036, Pakistan's three rivers, the Indus, Jhelum and Chenab, were delivering much less than half of their historic supply of water. In terms of water per person, Pakistan was only receiving 250 cubic metres a year: one-quarter of the United Nation's "critical" level.

India was also suffering severe problems with food supply by 2036. With the global temperature rise cutting into its existing productive capacity, it had only managed to expand domestic food production by 20 percent in the previous three decades, despite investing huge amounts in agriculture. Meanwhile, its population had grown from 1.1 billion to 1.5 billion. But India's plight was much less grave than Pakistan's, since only a small proportion of its agriculture depended on glacier-fed irrigation systems. Even in the eastern (Indian) Punjab, where grain production depended heavily on water from India's share of the Indus system, the three "Indian" rivers, the Sutlej, Ravi and Beas, relied much less on glacial melt for their flow and still delivered almost as much water as before. That, however, was precisely the fact that annoyed the Pakistanis so much.

"Annoyed" is too weak a word, really. The Pakistanis had always felt that the crude division of the waters embodied in the Indus Water Treaty had been skewed unfairly to India's advantage, given that India was using very little water for irrigation, even from the eastern rivers, at the time when the British Raj was originally divided between the two successor states in 1947. There was a chronic, and sometimes accurate, Pakistani complaint that the great powers tended to favour India when they mediated between the two countries because India, so much bigger and richer, was more important to their own long-range plans. Now, on top of that, came the bitter realization that the western rivers were failing Pakistan, while the eastern rivers still gave water to India.

There had been no overt military confrontation between the two countries since the Second Kargil Crisis of 2017, and even that had not gone beyond infantry and artillery clashes in Kashmir. The knowledge that the other side also possesses nuclear weapons instills a large degree of caution in even the most irreconcilable opponents. But almost ninety years after Partition, the mutual hostility between the two countries had not really faded, and the political temptation to scapegoat the wicked neighbour was ever-present. Pakistan had been in one of its interludes of civilian rule since 2018, but the helplessness of the elected governments in the face of the rapidly growing crisis of water supply (and therefore of food supply) steadily undermined their credibility.

In June 2035, the sixth military coup in the country's history brought a "Council of National Salvation" (CNS) to power. It differed from the previous five coups in two important respects. One was that it was carried out by relatively junior officers, mostly colonels, without the consent of their legal superiors: for the first time, the armed forces were not intervening as a disciplined whole. The other, closely connected to the first, was that it was not the usual bloodless affair. Generals were rounded up and, in a number of cases, when loyal troops tried to protect them, there were firefights in the streets. Some generals were subsequently executed, as were some politicians from previous governments whom the colonels felt had betrayed the national interest. It was early July before the CNS was fully in control of the country, and by then Pakistan was polarized as never before. The young colonels probably had majority support among an increasingly desperate population, but they certainly felt the need to solidify that support through prompt and dramatic action on the key questions of water and food.

In late July, they publicly demanded that India stop taking its small, treaty-mandated share of the water from the upper reaches of the Jhelum and Chenab rivers, and more importantly, that it give half the water from the eastern rivers to Pakistan. (Even if India were to grant their demands, it would only have solved about a third of Pakistan's water problem, but the young officers had no ideas about how to address the rest of it.) They also secretly instructed the directorate for Inter-Services Intelligence to send guerrillas into the Indian-controlled part of Kashmir, as it had done during a number of previous crises—but this time they had orders to attack the headworks on the rivers that flowed through the disputed province in order to raise the pressure on New Delhi.

The Indian government had been watching the growing desperation of Pakistan's leaders with concern for years, well aware that it could have grave implications for India, and the Ministry of External Affairs and the Indian armed forces had both gamed possible crises arising from it (though never together, of course). But neither India's generals nor its diplomats had foreseen such a radical and unpredictable regime coming to power in Islamabad and, over the ensuing months, there was much inconclusive debate in South Block about how to deal with it.

The guerrilla raid on the giant Bhakra Dam on the Sutlej River in March 2036 put the turbines in the north powerhouse out of operation for several months, but its main effect was to trigger the collapse of the coalition government in New Delhi. The new coalition that took its place in April included two hardline nationalist parties, and was pledged to put an end to the "terrorist" threat at any cost. The simplest way to do that, it reasoned, was to put unbearable pressure on the Pakistani sponsors of the attackers, and so it mobilized the Indian armed forces and moved a number of armoured formations up to the border with Pakistan.

There was no intention actually to attack Pakistan, of course, but the planners in New Delhi reckoned that the deployment of Indian armour near the frontier would force Islamabad to flood the very extensive network of "defence canals" that it had built at strategic locations along the border to prevent crossings by Indian armour and artillery, in the event of war. The water diverted to these canals would greatly exacerbate the existing water shortage in Pakistan and would put the Islamabad regime under great pressure to back down from the confrontation.

That was the plan, but while the young colonels in Islamabad did flood the defence canals as predicted, their political position at home was too precarious to let them back down gracefully. Foreign exchange to buy scarce and very expensive food on what was left of the international grain market had run out, food rationing had already been imposed in Pakistan's major cities, and they desperately needed a victory—so they issued an explicit threat to use nuclear weapons if India did not withdraw its troops from the border.

This was a potentially suicidal threat, since India had at least as many nuclear weapons as Pakistan and much better protected delivery systems. But the Council for National Salvation was operating on the familiar fallacy that the other side, in addition to being in the wrong, is morally inferior and will lose its nerve if pressed hard enough. What the Indians actually did was to start planning a non-nuclear pre-emptive strike to destroy Pakistan's nuclear delivery systems on the ground, using swarms of the small (thirty to fifty kilograms), heavily stealthed, highly accurate unmanned aerial vehicles that India had acquired as a result of its close defence relationship with the United States. A few Pakistani nuclear weapons would doubtless escape, but Indian air defences and anti-ballistic missile defences would deal with most of them and, at worst, India would take a couple of nuclear hits. Waiting for the "crazy colonels" to carry out their threat, it was argued, would lead to an infinitely worse outcome.

It should have set off alarm bells in New Delhi when the U.S. government started withdrawing all military and diplomatic personnel from India and urgently advising all American citizens to leave, but the Indian plan was set into motion at 3 a.m. local time on May 25, 2036. Well over half of Pakistan's nuclear weapons were destroyed before dawn on that day, by thousands of Indian UAVs homing in on their locations with relatively small packets of high explosives. As the crisis deepened, however, many of Pakistan's nuclear weapons had been redeployed from their usual locations, and Indian satellites had not picked up on all of the new locations. As soon as the Indian UAV strike got underway, the Pakistani colonels ordered instant nuclear strikes on all the bases in India that were believed to harbour nuclear weapons. In principle, the Pakistanis were still trying to avoid cities at this point, but their target list included places like Agra Air Force Station (long runways, big aircraft capable of launching lots of cruise missiles) and Bombay Dockyard (ships and submarines with cruise missiles), only a couple of kilometres from the city's heart.

By Day Three, even these crude distinctions among targets had been lost, and the remaining nuclear weapons (not many, by now) were mostly just aimed at population centres. The logic of this kind of conflict is inexorable: you cannot "win" a nuclear war, but if you want the survivors on your own side to have any future, you will try to ensure that they are not outnumbered by the survivors on the other side in the places that are important for their future. Besides, there really was no command and control left at this point: whoever still had a weapon, fired it at whatever target they had coordinates for— and everybody with a computer or even an atlas had coordinates for the cities.

On the sixth day, the nuclear war sort of petered out for lack of long-range weapons, and the de facto ceasefire became a formal arrangement a few days later, when provisional governments began to assert their authority over what was left of the two countries. Fatal casualties in India and Pakistan, including radiation victims who died in the first month after the war, were on the order of four to five hundred million, though exact figures will never be known. The Himalayan mountain wall protected China from most immediate fallout, but the prevailing winds delivered large doses of it across Bangladesh, Burma and northern Thailand, where some millions died as a result. There were not nearly enough weapons detonated to cause a "nuclear winter," but sufficient dust was boosted into the atmosphere to cause a cooling of about one degree Celsius in the Northern Hemisphere during the summer of 2036.

By the winter of 2036–37, columns of armoured vehicles, modified to provide some protection against radiation, were making their way back into the devastated zones on both sides of the old frontier. To avoid the hot spots, they used radiation maps plotted by UAVs. The aim was not so much to bring help to survivors who had not managed to flee—there were not enough resources for that—as to reassert control over the national territory. India had to divert a great deal of its effort to the northeast, however, as neo-Naxalite groups had taken advantage of the chaos to seize control of much of Orissa, Jharkand and Bihar states, and the suspension of armed patrols along the border fence had allowed millions of Bangladeshi economic refugees to pour into the states of Assam and Tripura. That distraction, plus the relatively greater distance to be traversed by Indian columns, allowed Pakistani forces to gain control of the state of Jammu and Kashmir, together with the headwaters of all but one of the rivers of the Indus system. So it was a Pakistani victory, in the end, albeit one that would have shocked even Pyrrhus.

CHAPTER FOUR
We Can Fix This . . .

Because these feedback mechanisms have now kicked in and they're going very rapidly, some people think it's too late already. Other people think we might have ten years — ten years to severely cut the carbon dioxide forcing function. This is an issue of first, the technology, and then, the will and the culture to do it. I think the technology is there . . .

The estimates are that in the U.S., if we went full-bore on conservation, we could reduce about 30 percent of our greenhouse-gas emissions. [For the rest,] there are four approaches which have the capacity to replace fossil fuels. Capacity is important, because an awful lot of the things that people are now espousing will not provide the requisite amounts of energy.

First and foremost, there are the biofuels. Not ethanol with sugar beets or corn; that's not good. Ethanol does not have the heating value, it requires too much energy to produce, it is not compatible with the existing infrastructures, and of course, there's not nearly enough of it. People are now talking about cellulosic biofuels, but they're still talking about plant stock grown on arable land with fresh water, and there's just not enough fresh water and arable land to produce enough biofuels to replace the petroleum.

So, therefore, the suggestion is to go to two other sources, which are algae and halophytes. On the shorelines of the Indian subcontinent they have had a brackish-water agriculture for many centuries using halophytes—salt-tolerant plants . . . No one is yet trying to do halophytes for biofuels on land using sea water irrigation except for Carl Hodges down in northern Mexico, but this is doable.

I've saved the best biofuel for last, and this is algae. Algae is 30 to 60 percent oil, which means that you can use very simple refining procedures to produce biodiesel. Also, algae is extremely productive: algae is capable of some twenty thousand gallons per acre-year in fuel, which is forty times the best that you'll get out of the best dry-land plants, be they cellulose on sweet-water arable land or halophyte salt-water agriculture. This factor forty is important, because the estimates are that if we wanted to produce enough biofuels on land to replace petroleum for U.S. use, we would have to use some 40 percent of the U.S. land mass. Using algae, you're down to less than 1 percent—and, of course, algae uses salt water.

The capacity is there, so biofuels will replace petroleum very rapidly, and the cost of the biofuels is estimated by the National Renewable Energy Laboratory in the U.S. at less than a dollar a gallon by the 2020 time frame . . . So, that's replacing petroleum. What do you do about the major issue, which is replacing the coal plants for the electricity?

Here one has to look at what's called base electrical load, which means 24/7 energy. What's possible here, first of all, is biofuels: you can burn them, because you paid the carbon dioxide price up front. The second approach is geothermal. On half of the land masses, if you drill down some two kilometres, you get two-hundred-degrees Celsius rock. If you drill down five kilometres, you get three-hundred-degrees Celsius rock. The capacity for drilling from the oil-well people is ten kilometres, so drilling a two-kilometre hole is not hard. Generally, the rock is fractured down there, or you can fracture it, so you drill a hole, you move over a bit, you drill another hole, you put water at a pressure of three thousand pounds per square inch down one hole, you get steam out the other, and you make a power plant. Compared to paying for a nuclear power plant, this has got to be less expensive.

Dennis Bushnell, chief scientist, NASA Langley Research
Center, in an interview with the author, February 2, 2008

AND ON, AND ON: solar thermal power plants that can concentrate the equivalent of sixty thousand suns on a boiler to run a turbine, nano-plastic solar photovoltaics, wind power, ocean tidal power . . . Dennis Bushnell is a man who is so enthusiastic about his job that, listening to him, you have to sort of lean into the wind a bit. It's also bracing to talk to him because he makes you see how easy this problem would be to solve if you just wrote off your sunk costs in old technologies and started with a clean sheet. We got into this fix by starting to burn fossil fuels on a large scale two hundred years ago, when there were no other technologies for producing large amounts of energy. Fair enough: you have to start somewhere, and at the time, that was our only option. Now there are lots of other technologies that can produce the energy we need, and all we have to do is stop burning the fossil fuels and use the new ones instead. As Dennis Bushnell enthusiastically explained to me:

> About everything I've discussed so far can be done, if we decide to do it, within ten to fifteen years. If we just change the approach, [forget] about the sunk investments in the existing fossil fuel infrastructure, and march off in the direction we need to go in to head off this really major disaster for the planet as well as the humans, we could possibly pull this off in a reasonable time frame.

I love Bushnell's approach, because I like living in a high-energy civilization, and I don't want to give it up. If it can be managed without causing a climate disaster, I would like everybody on the planet to live in wealthy societies that have the resources and the leisure to start looking after all their citizens and not just the top dogs. I would like the enterprise of science to flourish, and that is definitely only possible in a high-energy society. I would like to go on flying to distant places, too, and I would even wish for more people to have that same opportunity, even though I know that the more of us who travel, the more homogenized the destinations become. The human race, split and scattered and divided into ten thousand different and often warring societies by history and geography, is visibly moving towards some kind of global civilization in which we all have an equal stake, and I do not want that process to be aborted. So I want us to sail through this inevitable crisis about where our energy comes from and what it is doing to the climate with the least possible damage, and Dennis Bushnell has my vote.

However, I have to recognize that his proposal does involve closing down the entire oil and gas industry, and the coal industry too, within the next ten or fifteen years, and that this might pose a few political problems. People chatter gaily about "creative destruction" as the fundamental virtue of capitalism but, in the real world, it is remarkable how the "sunset industries" use their cash and their political clout to stop the sun from going down on them. Wartime-style mobilization and government controls might enable us to create alternative, non-fossil-carbon energy industries to meet all the demands of a high-energy civilization within ten or fifteen years, but the experience of Amory Lovins in this respect is instructive.

For thirty years, the most prolific originator of carbon-saving, climate-friendly technologies in the United States has been Amory Lovins, cofounder, chairman and chief scientist of the Rocky Mountain Institute in Colorado. It is fair to say that, if all his ideas had been put into practice across the American economy, the United States would now be producing less than half the carbon dioxide that it actually does. In reality, however, the U.S. automobile fleet still runs almost entirely on oil and gets some of the worst mileage figures in the world, and now biofuels are all the rage and nuclear power appears to be making a comeback. But Lovins is still confident that in the end people will choose the right technologies — the economically sensible technologies, as he quite rightly insists — and that this will solve the problem. Or at least, that it will solve enough of the problem, soon enough, to keep the show on the road.

> I start from the premise that we're making fifty-year changes which require relentless patience . . . It's perfectly true that most of the energy efficiency we could and should buy just to save money remains unbought. If we fully applied today's best techniques, we'd save over half the oil and gas in the U.S. and three-quarters of the electricity at an eighth of their cost. The reason so little is bought is that there are sixty or eighty specific market failures or obstacles or barriers — each of which can be turned into a business opportunity, if we really pay attention to barrier-busting — but most of us haven't really paid attention.

> The more that the concerns about price, climate and security converge, the more of us will pay attention and the more entrepreneurs will start helping us do so . . .

> Here's a strange-bedfellow story. The Pentagon is the world's single biggest buyer of oil, although it's only 0.4 percent of the world market. If it were an airline, it would be the first- or second-biggest airline in the United States. But most of its fuel is wasted, and it's starting to figure out that it's extremely expensive, in both blood and treasure, to deliver fuel to the planes, ships and land

vehicles that use the fuel, because they were designed as if delivery were free and invulnerable.

It's not. We have whole divisions of people hauling oil around and more divisions trying to guard them. About half the casualties in theatre are associated with convoys. Seventy percent of the tonnage being hauled is fuel, which is then largely wasted. So the Pentagon changed its policy in April 2007. In the future, it's going to require [the contractors bidding on new military vehicles, ships and aircraft to include in their estimates] the fully burdened cost of delivering the fuel. That means that they'll be valuing saved fuel enormously more than in the past, and the prime contractors will be competing over who can make the most efficient stuff.

Well, that's going to speed up how quickly we get those triple-efficiency civilian cars, trucks and planes, because remember that the last things that we got out of military R&D [research and development] included the Internet, the Global Positioning System, the jet engine industry and the microchip industry

Amory Lovins, cofounder, chairman and chief scientist,

Rocky Mountain Institute, in an interview with the author,

May 5, 2008

He might forgive the sixties phrase, because in some deeply submerged sense that is where he comes from culturally, but what Amory Lovins has actually undertaken is "the long march through the institutions." Don't preach; bargain. Push the agenda forward one deal at a time. Never make a frontal assault on the defenders of the old technologies and the old orthodoxy; just go round their flanks. He has accomplished a great deal, and yet so much more remains to be done. To stay with the military analogy, perhaps, in the end, you do have to make a decisive assault. If so, the first volunteer to lead the charge would be Lester Brown.

Les Brown is a hero. He has been in the trenches of the climate wars for thirty years, he lost control of Earthwatch, the organization he founded and made famous, in an internal coup a decade ago, and he still radiates a determined optimism about the ability of the human race to respond intelligently to the crisis that confronts it. His latest book, *Plan B 3.0*, published in 2008, is as comprehensive and rational an approach to decarbonizing the global economy as you could hope to find. Not all his calculations would work out in practice, because that never happens for anybody, but his entirely feasible proposals could probably deliver us 80 percent cuts in carbon dioxide emissions by 2020, if the political will were there. Moreover, nobody would be required to make great sacrifices or change their lifestyle drastically in order to reach this goal. And he unhesitatingly accepts the same deadline as Dennis Bushnell.

> [There's no point in] asking the politicians "What do you think we can do?" because they'll say "We're for 80 percent cuts by 2050." In my book, the game's going to be over long before 2050. So we ask the question: how rapidly can we cut carbon emissions if civilization is at stake here? And it's in that context that we concluded that we should be cutting carbon emissions by 80 percent by 2020, twelve years from now.
>
> That would stabilize atmospheric carbon dioxide concentrations at just a hair under 400 parts per million—we're at 382 now—and then you'd be in a position to begin bringing carbon dioxide concentrations down. Now, most people look at these numbers and can't get their arms around them. This is big-time change coming very fast.
>
> In the 2020 energy economy, wind will be supplying 40 percent of world electricity needs. That means installing a million and a half wind turbines around the world that are on average two megawatts each: that is three

million megawatts of wind generating capacity. It sounds like a lot, but this is over the next dozen years. We build sixty-five million cars [around the world] every year, so it's not a big deal. We could produce these wind turbines for the entire world simply by opening the closed automobile assembly plants in the United States. That's all we need to get 40 percent of our electricity from wind by 2020 . . .

The other interesting example is building solar-thermal power plants: mirrors that concentrate sunlight on a container with a liquid that produces steam and drives a turbine. There are now solar-thermal power plants in operation or in planning in California, Nevada, Florida, Spain and Algeria. Algeria [is] planning six thousand megawatts of solar-thermal power plants in the desert, and they're going to cable that electricity to Europe, under the Mediterranean. They know their oil reserves will be gone one day, but they plan to replace those exports with exports of solar electricity.

I mention these examples to give a sense that things are beginning to happen on a scale that's commensurate with the nature of the challenges and threats that we are facing.

—Lester Brown, founder, Earthwatch, founder and
president, Earth Policy Institute, in an interview with the author,
January 30, 2008

Brown is not politically naive. He grounds his hope in existing technology, and in the expansion of projects for cutting emissions or generating energy from non-carbon sources that are already being put into practice somewhere or other. He puts little faith in international diplomatic efforts like the negotiations for a follow-on Kyoto accord, and a lot more in grassroots initiatives like the Sierra Club's campaign to get all new coal-fired power stations banned in the United States. In *Plan B* 3.0, he talks about a race between "human" tipping points where the political will to do something about global warming finally manifests itself, and "natural" tipping points where climate changes move beyond the human ability to reverse them. His optimism is not a character flaw; it is an operational necessity, and one that clearly exacts a toll on his energy every day. But it is a *reasoned* optimism, in the sense that everything he says is grounded in current technologies.

If one were to offer a small criticism, it would be that sometimes, in his enthusiasm, Brown paints with a very broad brush. For example, the wind turbines that will allegedly produce three million megawatts of power and supply 40 percent of the world's energy needs in 2020 put out two megawatts each, but only when they are running at maximum revs in a good wind. That's only a small proportion of the time: most of the time, they put out a lot less power, so we would really need around five million of them. They'd have to be linked up in immense grids across continents and oceans in order to minimize the problem of variable winds, and they'd still need back up from some other power source for those times when the wind isn't blowing strongly enough to meet demand. But these are mere quibbles: if the world can build sixty-five million cars a year, it can build five million wind turbines in ten years.

Not as an antidote to Lester Brown, but as a necessary supplement, there is George Monbiot, the master of gritty realism. In his 2007 book, *Heat*, he set himself the task of figuring out how Britain, the oldest industrial nation and the first to emit really large amounts of carbon dioxide and other greenhouse gases, could become a carbon-neutral society. Or at least something close to that: the actual target he sets himself is a 90 percent cut in British emissions by 2030.

The great virtue of Monbiot's work is that he actually crunches the numbers. He is diligent and ruthlessly honest in his treatment of the data, which leads him to some interesting discoveries. For example, he concludes that far from dismantling the national electricity grid in favour of local co-generation schemes that produce both electricity and heat, as the fashion-able wisdom has it, a low-emissions Britain might actually need to expand the grid (using high-voltage DC cables, of course) in order to connect wind farms out at sea with all the customers who need electricity but do not need heat. He also discovers, over and over again, that the "facts" about the true costs of all the rival forms of power generation, whether coal, gas, solar, wind or nuclear, are so unreliable and divergent as to be close to useless. Depending on which side of the argument you are on, you just pick your data to suit your case.

Monbiot actually took the trouble to understand how big power grids work, so he couldn't dismiss all the complexities with the usual airy wave about how we'll do it all with "renew-ables." The central and deeply intractable fact about electricity is that it must be generated at the very moment when it is used. This poses a major problem for those in charge of supplying a society with electricity, since in a country like Britain the demand can vary as much as threefold from a mild weekend day in summertime (twenty gigawatts) to a cold midwinter evening (sixty gigawatts). And there are occasions when it varies by almost three gigawatts within a minute, as it did when almost the entire British population got up and put on the kettle after the penalty shootout in the World Cup semifinal in 1990.

> Because our televisions and our lights were burning already, the system was straining even before these extra demands were made. Someone had to ensure that the extra power was found at the very moment it was required, and not just *some* extra power, but a quantity matched exactly to the demand . . .
>
> To respond to these fluctuations in demand, [the electricity companies] need constantly to be bringing power plants in and out of production. The "baseload"— the 20 GW we use all the time—is supplied by nuclear reactors and large gas-powered plants. As more electricity

is required, coal-burning stations and smaller generators are brought online. Some plants will be kept out of production all through the summer, and fired up as demand rises in the winter. Some are held in "spinning reserve": operating, inefficiently, at just part of their load, but ready to be brought up to full power in seconds. A cable between France and the United Kingdom allows us to import 2 GW of electricity when we fall short . . .

Most dramatically, the UK has three "pumped storage" plants, which provide our only cost-effective means of storing power. They each consist of two reservoirs, one at the top of a mountain, the other near the bottom. When electricity is cheap, which means when demand is low, it is used to pump water from the bottom reservoir to the top one. When there is a requirement for a sudden surge in production, the gates of the top reservoir are opened, and the water pours through turbines back down to the bottom.

The pumped storage plant at Dinorwig, in north Wales, can produce 1.7 GW of power for five hours. It responds within fifteen seconds. I like to picture a man in a booth with his television on. As the match draws to an end, the phone rings. "It's the last penalty. Open the gates." He pulls down a great red lever, and the water roars out of the upper reservoir just as the ball thumps into the corner of the net. I'm sure that in reality it's all done automatically.

—George Monbiot,
Heat: How We Can Stop the Planet Burning, 2007

Try though he does, Monbiot cannot plausibly imagine a national electricity grid that draws more than 50 percent of its power from "renewables" (mainly wind, in Britain's case). You can try to smooth out the fluctuations in power coming from wind farms by spreading them across a large number of widely separated locations, but the fact remains that wind speeds fluctuate within a wide range at every site.

The managers of the big grids, who already have to cope with wildly fluctuating levels of demand, are now being asked to deal with uncontrollably fluctuating levels of supply as well, and they can only go so far. It hasn't come up much in public yet, because only Germany and Denmark have approached even 20 percent renewables in their electricity-generation mix, but most grid managers are very unhappy about going beyond that level of renewables in the system. With better programmes for predicting short-term wind fluctuations they might be persuaded eventually to go to 50 percent, but beyond that, Monbiot concludes in *Heat*, you simply cannot go unless you are prepared to accept periodic collapses of the entire grid when the wind drops. And you can't actually "replace" half of the existing carbon-fuelled power stations with wind farms, either. You have to keep most of the old coal and gas-fired stations, too, in order to provide the needed power when the renewables cannot.

Most of the time those reserve power stations won't be running at full load, and many will not be fired up at all, but even the 50 percent of your electricity that comes from renewables will still involve burning some coal or gas. As for the other half of your supply, it will still be coming from base-load generators that are either nuclear or carbon-fired. So, no 90 percent cuts in emissions in the electricity sector.

Monbiot plows on regardless, making "best guesses" when all the data lie and never losing sight of his goal. He rigorously confines himself to off-the-shelf technologies that are available today, with one single exception. There is nothing else like it in the literature, because Monbiot really engages with the recalcitrant realities of how a late industrial society might cut its emissions by 90 percent while continuing to live in more or less the style that it is accustomed to. (A necessary condition of the argument, since the likelihood of an entire society being persuaded to abandon its whole lifestyle in order to avoid a hypothetical disaster that is several decades in the future is very small.) For all his ingenuity, however, he can't save cheap mass air travel from the chop: goodbye, long weekends in Marrakesh and shopping trips to New York.

But relying only upon existing technologies and assuming a far greater degree of political will than exists in Britain at the moment, Monbiot does manage to reach the finish line. The Britain of 2030 that has cut its emissions by 90 percent would be recognizably the same country as today, except that it draws most of its power from different sources (including the power that drives its motor vehicles), lives in much better insulated houses, and travels mainly by rail and bus. It can be done, and it can be done relatively quickly. So why hasn't it been done already? Why is this game going into overtime?

This generation has altered the composition of the atmosphere on a global scale through radioactive materials and a steady increase in carbon dioxide from the burning of fossil fuels.

—President Lyndon B. Johnson, 1965

The fact that carbon dioxide is a greenhouse gas was established in the nineteenth century by Irish scientist John Tyndall. In the early twentieth century, the Swedish scientist Svante Arrhenius concluded that the amounts of fossil fuel being burned by human civilization could ultimately warm the climate—a warming, albeit very slight, that was actually observed by British engineer Guy Callendar in the 1930s— but the phenomenon got little attention until the 1960s. By then, the observations on the trend of carbon dioxide concentrations in the atmosphere that Charles David Keeling began taking at Mauna Loa Observatory in the 1950s were producing a different volume and quality of evidence, and in 1965, the Environmental Pollution Board of the President's Science Advisory Committee in Washington warned Lyndon Johnson that human activities "will modify the heat balance of the atmosphere to such an extent that marked changes in climate . . . could occur."

Lyndon Johnson dutifully incorporated it into a speech (above), but that should not be taken to mean that he planned to do anything about it. He had the Vietnam War, the civil rights movement, the Cold War and a few other priorities on his plate at the time, and besides, in another corner of the forest, a different group of scientists was worrying aloud about the return of the ice age. It didn't seem very urgent to sort out which lot were right. After all, the concentration of carbon dioxide in the atmosphere was only just over 320 parts per million, a mere forty parts per million higher than it had been at the beginning of the Industrial Revolution in the late 1700s.

Slowly the evidence piled up that climate change was a serious danger, although the subject surfaced only intermittently outside scientific circles. By 1979, the JASON Committee, a group of scientists with high-level security clearances who gathered annually to offer advice to the U.S. government, predicted that carbon dioxide concentrations in the atmosphere might double by 2035, resulting in a rise of two to three degrees Celsius in average global temperature, and polar warming of as much as ten to twelve degrees. The National Academy of Sciences, which the Carter White House asked for a second opinion, set up a committee that concurred with the JASON conclusion: "If carbon dioxide continues to increase, [we] find no reason to doubt that climate changes will result, and no reason to believe that these changes will be negligible." But the Iranian Revolution had just happened and President Jimmy Carter was preoccupied with the hostage crisis, the second oil shock, and a few other things. And the carbon dioxide concentration was still slightly under 340 parts per million.

You can't really fault these people for treating climate change as a distant and largely hypothetical threat. They were living in a complicated and frightening time, and the whole notion that human activities might alter the balance of the entire planet was still a quite new concept. Even those who had grasped that the temperature would rise if carbon dioxide emissions continued had little notion of how unpleasant the consequences of that might be for human civilization, and besides they had few technological alternatives to burning fossil fuels except nuclear energy. In fact, the sudden burst of action on the climate front that lasted from the mid-1980s to the mid-1990s was an impressively prompt and coherent response to a problem whose full dimensions had only just been understood.

It was accomplished by a relatively small number of individuals, perhaps because at that time there was not yet an organized industrial lobby seeking to discredit the science. For example, Sir Crispin Tickell, then a senior British diplomat, had devoted a sabbatical year in the early 1970s to studying the nascent science of climate change (although he was a historian by training) and had subsequently written a book about both the environmental and the political implications of global warming. As the evidence mounted in the early 1980s that something was seriously amiss, he was able to influence the position of the British government with astonishing ease.

> In the early 1980s, I had the opportunity of talking to [Prime Minister] Margaret Thatcher [about climate change], and found that she was very sympathetic, and from that point developed a great many other things, including the inscription of climate change on the agenda of the G7 summit of 1984 . . . She was, at that time, the only scientific member of the British government . . . We had, later on, a meeting of senior ministers in Downing Street at which this dimension was fully explored. Jim Lovelock came to that, I was one of the

speakers, and we explained it all, how important it was to politics. And, of course, Margaret Thatcher herself made a very famous speech to the Royal Society on this subject which I modestly helped with, and that again set the scientific dimension of a big political issue very loud and clear.

—Sir Crispin Tickell, in an interview with the author,

July 3, 2008

The American scientist James Hansen gave a famous speech to Congress in 1988 that had a comparable impact on American politics, and after that, things moved very fast. The Intergovernmental Panel on Climate Change was established in 1988, by the World Meteorological Organization (WMO) and the United Nations Environment Programme (UNEP), with the task of monitoring the changes and predicting their future course, and almost all the countries of the world became members. By 1992, the Earth Summit in Rio, formally the United Nations Conference on Environment and Development, brought 172 countries together and produced the UN Framework Convention on Climate Change, whose stated objective is "to achieve stabilization of greenhouse gas concentrations in the atmosphere at a low enough level to prevent dangerous anthropogenic interference with the climate system."

Everybody signed up, and the U.S. Congress ratified the treaty in only three months. Soon a national greenhouse-gas inventory was created—an accounting of greenhouse-gas emissions and removals for each country—with reports regularly submitted to UNFCCC headquarters in Bonn, Germany. That provided a means of monitoring any cuts in emissions that might subsequently be agreed by the signatories of the treaty. However, the member countries had not yet accepted a legal obligation to do anything specific—and when they finally had to do that, at the Kyoto meeting in 1997, the wheels almost came off.

It was the place where the commitment to contain climate change that the countries had undertaken at the Earth Summit five years before would finally be translated into actions. For the first time, the countries present were being asked to accept specific curbs on their emissions that would have direct impacts on their economies—and that was alarming, because nobody could be sure where the open-ended obligation "to achieve stabilization of greenhouse gas concentrations in the atmosphere at a low enough level to prevent dangerous anthropogenic interference with the climate system" might lead them in the end. So the haggling began, with most countries seeking to minimize their cuts and the poorer ones trying to avoid them entirely. What else would you expect? So it came down to salami tactics in the end. The principle that all countries are jointly responsible for the level of carbon dioxide in the atmosphere and for the fate of the climate was preserved—but this agreement was achieved by keeping the actual cuts quite small, and exempting the poor from them entirely. We'll secure the principle first, and work on the harder details later.

A great deal of nonsense has been talked about the Kyoto Protocol of 1997. Its more rabid critics have shamelessly worked both sides of the street, denouncing the treaty as irrelevant because climate change is a myth—and simultaneously complaining that the emissions cuts decreed by the Kyoto treaty are so modest that they will only cut global warming by one ten-thousandth of a degree in the year 2552 (or words to that effect). This is a bit like the classic boarding-school complaint—"The food here is disgusting, and the portions are too small!"—but the critics do have a point: the reductions in greenhouse-gas emissions mandated by Kyoto really were too small, and restricted to too few countries, to have a significant impact in slowing global warming. They were determined not by what was needed to stop global warming, but by what the countries present could be persuaded to accept.

Three things need to be borne in mind about the negotiations in Kyoto in 1997. At that time, it was not clear that global warming would move as fast as subsequent models have predicted (let alone as fast as some recent events suggest). Neither did most of the participants realize in 1997 that the recent ten-year growth spurt in China and the even more recent acceleration in India's growth rate were not transient events, but would both continue into the future at an even faster pace. Both countries were actually on course to become major emitters of greenhouse gases in their own right, but this was a very new and contentious idea in 1997, so leaving them without obligations to curb their emissions in the first round of negotiations was seen as relatively cost-free. And finally, there was already concern that the United States, the biggest emitter by far, producing one-quarter of the world's man-made greenhouse gases with only one-twentieth of the world's population, would not ratify the treaty if it demanded significant sacrifices from the U.S. The Clinton White House might be in favour, and Vice-President Al Gore put a lot of effort into the negotiations, but the Senate would probably kill any treaty that placed serious obligations on the United States.

So the treaty that emerged in Kyoto was an unambitious little thing, requiring the industrialized countries to cut their emissions by a very modest 5–7 percent by 2012 (from a 1990 baseline) and imposing no constraints on the emissions of developing countries, including China and India. But it was far from a waste of time, and until the end of the twentieth century, there was still reason to believe that the response would match the scale of the problem.

> It was actually a quite remarkable rate of progress, given the fact that you really couldn't see many effects of this global warming. It was a theory that we had some confirmation for, and we knew what we were talking about, but it was quite impressive. Already within four years

[after the creation off the IPCC in 1988] we had the Framework Convention on Climate Change, and then the Kyoto Protocol [in 1988]. But what happened was, as we all know, that the U.S. sabotaged the effectiveness of the Kyoto Protocol by not signing on. Without the biggest polluter by far, and without the biggest economy, Kyoto could not be very effective.

It would have been much more effective in reducing the developing country emissions because of the Clean Development Mechanism that was part of it . . . There would have been a lot more incentive to get moving on the renewable energies and clean energies. But why did the U.S. take that position?

Well, it was because the industries had more influence on our government than the public good. The interest of a small number of people, the influence of money in Washington, and the fact that so many congressmen and administration are influenced by the fossil-fuel industry, which is huge. That's why we need to try to draw attention to [the activities of that lobby], because otherwise they may continue to muddle the story enough that we fail to get the strong and rapid actions that we now need. Because now we've used up the time—we're at the hairy edge right now. We have to make rapid changes.

—James Hansen, director, NASA Goddard Space Studies
Center, in an interview with the author, June 28, 2008

U.S. President Bill Clinton never sent the Kyoto treaty to Congress for ratification, because he knew that it had zero chance of ratification: by now the fossil-fuel lobby had bought up enough members of Congress to ensure that. Shortly after President George W. Bush entered office in 2001, he declared that the United States was withdrawing from the treaty entirely—and the next eight years of international negotiations on climate change were characterized by official American obstructionism that often verged on wrecking tactics. Somehow or other, the issue of climate change had got caught up in the "culture wars" in the United States, and the Bush administration was on the side that denied it was happening.

It is not immediately obvious why the reality or otherwise of climate change, essentially a scientific question, should have become such an intensely emotional issue in the United States. It did not become a left-right struggle elsewhere (except in Australia and Canada, which were carried along in the cultural wake of the United States), and conservatives like former British prime minister Margaret Thatcher and Germany's Chancellor Angela Merkel could be found in the front rank of climate activists. Even allowing for the unusually ideological character of American politics, why did this particular issue become a badge of ideological allegiance in the United States?

The fact that there was a campaign of denial funded by the usual suspects in the American oil, coal and automobile industries (and partly run by the same people who had previously conducted the tobacco industry's campaign to cast doubt on the evidence that smoking causes fatal diseases) does not explain the emotion that Americans invested in the issue. After all, smoking never became a pure left-right issue in the same way. The passionate commitment to the cause that is evident in American blogs on climate issues, especially on the part of the deniers, is quite disproportionate to the impact that the topic has on the real lives of the participants. Sociologist Donald Braman, an associate professor of law at George Washington University Law School, recently conducted a large experiment that may explain why people reacted as they did. I spoke to him on February 5, 2008:

> We provided people with a report from a fictional scientific organization that came out contemporaneously with the large international report by the [IPCC] scientists all over the globe who essentially were saying to people:

"Look, the issue is settled. Global warming is happening. Humans are having an influence on it, and the question is how severe and how detrimental to our welfare that is." Using that, we created a fictional report that said: "It's happening, humans are responsible for it, and it's going to have these severe consequences. The remedy for it is . . ." And then one remedy that we proposed was tighter regulation of industry and pollution: a typical progressive, egalitarian, communitarian response to the problem of global warming. Another possible solution was deregulation of the nuclear industry to allow for the development of that kind of alternative energy at no cost to the atmosphere.

The responses were fairly dramatic. When you provide the conservative or individualistic-type folks with the condition that says we want more regulation of pollution, they see red. This is a disaster. They despise the suggestion that that is a solution, but then moreover, if you ask them how severe they think the global warming problem is, they say: not very severe at all. First, we don't really think it's happening. Secondly, if it's happening, it's not because humans are involved. And thirdly, if there are consequences, they're likely to be mixed. Some of them will be good, some of them will be bad, it's hard to say on balance that this is something that's terrible or to be avoided.

But show them the solution of deregulated nuclear power, and all of a sudden global warming is a real problem, and we need to deal with it now. Not only is it a threat to us very soon, but humans are contributing to it, and sure enough the consequences are going to be dire. [One solution] resonates with their preferred vision of how society should work: private orderings, deregulation, scientific knowhow overcoming the threat of environmental harm. [The other solution] is the polar opposite from their perspective: increased regulation, clamping down on private enterprise. And their perception of the risks that are associated [with climate change]—risks that shouldn't have anything to do with either of the solutions—really fluctuate quite a bit.

Interestingly, it also turns out to be the case that when you propose nuclear power as a solution to global warming, progressives start to fear nuclear power . . . much less. The progressives, the egalitarians and environmentalists who used to say: "Nuclear power is really dangerous, and I wouldn't want to live near a nuclear power plant. It ought to be highly regulated . . ." When they see it in the context of global warming and as a potential solution, they start to see the risks of nuclear power differently.

In less ideological societies, climate change could be treated as a more or less neutral fact. Given the ferocity of the culture wars in the United States during the Clinton and Bush administrations, global warming was bound to become a highly contentious "values" issue in the United States, rather than a scientific one. This was a major misfortune, since the Bush administration, whose international position on climate change was shaped by its role as a partisan in those domestic ideological struggles, certainly bore a large share of the responsibility for the lost decade in international action on the global-warming agenda. However, it is also the case that some other countries used American foot-dragging as an excuse not to pursue a more ambitious policy themselves. We have yet to know how they will act if they lose that political cover due to a change in American climate policy.

We may find out quite soon, however, since the period of time covered by the Kyoto Protocol is now drawing to an end, and the parties to the treaty have until late 2009 to decide on a follow-on deal that determines what further cuts will be needed in the period after 2012. Membership in the post-Kyoto negotiating group is now almost universal, with even Australia having signed up after the 2007 election removed prime minister John Howard, an inveterate climate change denier, from office. Under Prime Minister Stephen Harper, Canada has hinted at leaving the treaty and has ceased to work towards fulfilling its existing commitments, but it will probably be forced back into line after the inauguration of the next U.S. president in January 2009 (since the presidential candidates of both major parties in the U.S. are committed to action on climate change). It is even possible that the United States itself will sign up to the Kyoto Protocol or its successor in the next few years, although that is by no means guaranteed. But will the next round of negotiations really yield a dramatically different result?

> HANS-JOACHIM SCHELLNHUBER: I don't want to be pessimistic, but there are two possibilities. We might get a good post-Kyoto agreement by 2010 or so, which would mean that developed countries would have a road map that would include [a commitment to 40 percent emission reductions by 2030]. That's still a possibility. China, India and so on would have soft targets, but only for the commitment period until 2020 . . . [Then we would revise the whole thing in 2015, 2016,] so that they could enter in the third commitment period after 2020 to crisp numerical targets. In a sense, it would be a training camp for developing countries between 2012 and 2020. It would not ensure that we would get this 40 percent reduction by 2030, but at least we would be on track.
>
> If that does not happen, there is still the possibility that you would have a coalition of the willing, in the sense that just the most advanced countries—say Germany, U.K., Netherlands, Sweden, United States, hopefully Australia, where we had this sea change already, Japan—would just form a club of countries who want to become extremely energy-efficient, extremely

energy-independent, decarbonizing their societies, and do it through a patchy but ever evolving global-emissions trading system. Then market forces as well as political willpower would drive the whole thing. So we wouldn't get global targets, but they would simply emerge from the bottom, if you like. That would be the alternative. I think either way we will see something.

GD: Does that alternative approach keep us clear of the 450 parts per million threshold?

HANS-JOACHIM SCHELLNHUBER: Only if this is just a training camp. After 2020, clearly, you need to engage all the major industries in the world. I still think that if we are able, for a certain period of time, to demonstrate in a credible way that the most developed countries are on track, are willing and able to do this decarbonization, then this will become something like an infectious disease. If you see that you can lead a good life and nevertheless decarbonize your society, I'm sure the Chinese would be happy to copy that. But it would all depend on the success, on the shining demonstration of that period. Then I still think the 450 line could be held.

We might have a little bit of help from our dirty friends, namely the aerosols in the atmosphere. You know that they have this global dimming, masking effect. The aerosols might just help us over that period of time. It's almost ironic: without these aerosols we probably would have already much higher global warming. So it may turn out that if we would do a very subtle management of aerosols by sulphur filtering in China, India and so on, in line with carbon dioxide reduction, we might still save the day, but it may become a very tricky game, actually. So, the answer to your question is a definite maybe.

—Hans-Joachim Schellnhuber, director,
Potsdam Institute for Climate Change Research,
in an interview with the author, March 25, 2008

My interview with Hans-Joachim Schellnhuber was one of those interviews where you have to listen very carefully, but if you do, his answers will tell you a lot. Schellnhuber is the leading climate scientist in Germany, the major industrial country that has done more than any other to cut its greenhouse-gas emissions: it is officially committed to making 40 percent emission cuts by 2020. He is the climate-change adviser to Chancellor Angela Merkel, the only leader of a G8 country who is a fully qualified scientist: as Schellnhuber says, "we can talk in equations." Nobody could be more serious about getting emissions down urgently—and what he really expects to happen, on the evidence of the above interview, is the following.

He is not optimistic about an international agreement on a post-Kyoto treaty, and even if it does occur, he does not expect the rapidly developing Asian countries, where emissions are growing fastest, to accept hard targets before 2020. That would not ensure that the world gets to 40 percent cuts in greenhouse-gas emissions by 2030, "but at least we would be on track."

Alternatively, if there is no post-Kyoto deal, the rich industrial countries of the West-plus-Japan may just voluntarily create their own emissions trading club. That initiative will enable them to decarbonize their own societies so successfully that they create a shining example for the rest of the world, which will ultimately feel compelled to copy the example in order to attain the same level of prosperity and energy independence.

Since it would take at least a decade for the "shining demonstration" of the West to win everybody else over, that would mean another decade of rising emissions in the rest of the world and, at best, a gentle decline in overall global emissions during that period. Carbon dioxide emissions are cumulative, so how does Schellnhuber hope to hold the 450 parts per million line that is still officially the European Union's "never exceed" ceiling for carbon dioxide concentration in the atmosphere?

You almost have to decode what he says next, but he is actually pinning his hopes on "global dimming." He clearly does think that we will exceed the 450 parts per million threshold, at least for a while, but he hopes that the new coal-fired power plants and smokestack industries of Asia will pour enough aerosol pollution into the atmosphere, especially sulphate particles, that they compensate for the loss of polluting particles as Western emitters clean up their act. Provided there is still a reasonably dense shroud of pollution in the atmosphere, enough sunlight will be reflected back into space that the temperature may stay fairly stable, even if the carbon dioxide concentration bounces up to 500 parts per million or so for a while. And, if it doesn't get too hot, we may not trigger any feedback mechanisms. Then, once the Asians have started to clean up their act too, he hopes that a "very subtle management of aerosols" could keep their sulphate emissions high enough to counterbalance the impact of the higher carbon dioxide concentration, until the carbon dioxide level finally drops back down to 450 parts per million sometime later in the century. Which would be, as he admits, "a very tricky game."

A lot of time has been wasted, and it looks like some more will be, too, but this is how human politics usually works. We got off to a remarkably good start in dealing with global warming, and then we were derailed by the kind of sabotage that you can always expect when change threatens vested interests. Now there is a serious effort underway to get the train back on the tracks, and it may succeed. We just have to hope that climate change will move slowly enough to accommodate our political habits.

SCENARIO FIVE:
A HAPPY TALE

Today we witness a very great change for hydrocarbons. The [oil price is already] very high, and we think it will reach [US]$250 a barrel.

Alexei Miller, CEO of the Russian oil and gas giant
Gazprom, June 10, 2008

THE OIL PRICE DID NOT ACTUALLY REACH US$250 a barrel until August 2011, after a brief detour back down to below US$100, but the response in the main oil-importing countries was already well underway. In some ways it resembled the reaction to the previous oil-price peak in the 1970s and early 1980s, when conservation measures, such as lowered speed limits and more fuel-efficient cars, were the main tools used by the customers to reduce consumption and get the price back down. In the new crisis, however, conservation alone was not enough: the steadily rising demand for oil in the emerging industrialized economies of South and East Asia ensured that overall global demand continued to grow despite better mileage figures for American cars and trucks.

The desire to end dependence on imported oil had already driven previous U.S. administrations to mandate and heavily subsidize a switch to biofuels. It was technologically premature, in the sense that growing "industrial corn" for conversion into biofuel, though it made some Iowa farmers very prosperous indeed, never had the potential to make more than a marginal inroad into the U.S. dependence on imported oil. Growing enough biofuel from corn to supply the needs of the whole U.S. vehicle fleet would have required about half the land mass of the continental United States. But the idea of bio-fuels had gained credibility, and the prospect of a permanent haemorrhage in the U.S. balance of payments to pay for imported oil at hugely inflated prices created the political conditions for a leap straight into third-generation biofuels.

The Fuel Independence Act of 2012 provided for a gradual withdrawal of existing federal subsidies for first-generation bio-fuels over five years and redirected the funds to massive projects for growing oil-rich algae. Further funds were voted for intensive research into the most suitable "halophytes" (salt-tolerant plants) for growing in desert areas with seawater irrigation, and into commercializing the processes for converting carbon dioxide captured from power plants and directly from the air into octane fuel. The expected political backlash from the farmers whose subsidies were being withdrawn was blunted by the high prices and rising global demand for food: almost all of the land that had been diverted into growing "industrial corn" continued to be profitable under different crops.

Since the new fuels required no major transformation of either the distribution system or the vehicles themselves, by 2014 the proportion of biofuels in the U.S. fuel mix had passed 15 percent and was rising at 4 percent a year. Mexico, which recognized the market opportunity, plunged into growing oil-rich halophytes in its extensive desert coastal areas and became a major biofuel exporter to the United States. Total U.S. oil consumption dropped steadily, and by 2018, the proportion imported from overseas had dropped from two-thirds a decade before to only one-third. Plans for further expansion in the Alberta tar sands had already been abandoned, and there was confident talk of a coming time when oil would never be burned as fuel. Instead, the remaining supply would be reserved for use in the petrochemical industry, to make products like fertilizers, pesticides, plastic, and synthetic rubber.

The first countries to follow the American example were China and India, both of which were facing huge foreign exchange problems as the cost of their oil imports soared, and by 2015, oil imports were also dropping steeply in those countries. The Europeans took a bit longer to switch, and when they did, they concentrated much more heavily on making octane fuel by doping carbon dioxide with hydrogen. Indeed, the European Union deal with North African countries that saw huge amounts of solar-generating capacity built with EU money in the Sahara in order to provide Europe with electricity (delivered by high-voltage DC cables running under the Mediterranean) was partly motivated by the need for the large amounts of power used to split hydrogen from sea water.

All of these measures, driven by the high cost of oil, only addressed the one-fifth of human greenhouse-gas emissions that were directly connected to burning petroleum-based fuels in vehicles. The other four-fifths of emissions, from electricity generation, deforestation and agriculture, were hardly affected. But a series of regional climate-related calamities—the storm surge that inundated most of the Nile Delta and made ten million Egyptians homeless in 2011, the summer-long heat wave that caused at least seventy-five thousand deaths in the American Midwest in 2013, and the catastrophic floods on the Yangtze, Mekong, Salween and Brahmaputra rivers in 2014—served to mobilize public opinion on the issue of climate change, not just in the most severely affected countries but all around the world.

By 2015, the largely toothless Copenhagen Protocol of 2010, the successor to the entirely toothless Kyoto Protocol, was being overtaken by the "Zero-2030" movement. The goal of achieving zero carbon emissions worldwide by 2030, originally popularized by an Indian environmental group, captured the global imagination, and in the next few years, national governments adopted previously unthinkable programmes for not just cutting but entirely eliminating the burning of fossil fuels. The dream of "carbon capture and sequestration" at an economically viable cost was officially abandoned in most major emitting countries, and governments began the hard work of devising energy systems that did not rely mainly on carbon-based fuels.

Immediately available, off-the-shelf technologies like wind, wave, tidal, solar and nuclear power were naturally much in favour, because those were solutions that large amounts of money could be spent on right away. But "deep geothermal" pilot projects proliferated in those parts of the world where there were vast tracts of hot rock only a few kilometres beneath the surface, and very large amounts of money were poured into research on more exotic proposals for generating energy without emitting greenhouse gases.

There's a very interesting collection of proposals which we've called wild cards. The first one of these is "p-boron fusion." This is similar to the deuterium-helium 3 cycle . . . [but] this is aneutronic fusion, [which] produces protons instead of neutrons. [It's] much safer in terms of radiation hazards and so forth. There was a recent twelve-year Office of Naval Research study on this, which showed that it was much better than the conventional deuterium-helium fusion approaches . . . and is worth working on.

The second [wild card] is an update on what was called "cold fusion." Cold fusion, when it was first discovered, had great replication problems and absolutely no theoretical understanding. Everybody said "you people are smoking something." And so it went into disrepute — except that human nature is such that there are curious, stubborn people, and around the world . . . there have been vast amounts of experimental data acquired over the ensuing couple of decades showing that yea, verily, it's real. What's interesting is that last year there was a theory produced by Lewis Larsen and Allan Widom, and what they said was that this is electro-weak interactions on surfaces, which [are] fully explainable by the standard model of quantum theory. And they can explain all the idiosyncrasies in the data, all the replication issues, and they are busily trying to nail down the intellectual property, as you can imagine, and then try[ing] to figure out how to produce much higher-quality heat, and they are making great progress. So, there is the outlook for that.

The third [wild card], which is even farther out and has its own giggle-factor associated with it, is "zero-point energy." Zero-point energy is the zeroth quantum state. Zero-point energy is real, [it] is predicted out of quantum electrodynamics; the theory has been checked out to some fourteen decimal places, more than almost any other physics theory. If you integrate the energy in zero-point energy, the zeroth quantum state, out to the Planck scale, what you get is a *huge* amount of energy which is not observed cosmologically. Yes, there is dark energy; yes, the dark energy is probably associated with zero-point energy, but there's an extremely huge mismatch between the two, so we don't really know what its actual nature is. However, that hasn't stopped some very clever people, including the chief scientist out at Lockheed [Martin] Sunnyvale and some other very good people, from postulating seven different approaches to tapping zero-point energy. And if any one of those works out, and the fact that there's seven means, well, maybe one of them might, that could change this whole energy picture yet again.

But you don't have to count on any of that. We have ways forward which, in fact, will work without . . . terribly time-consuming or expensive further technological developments. It's simply a matter of giving up our current teddy bears, which we love to clutch, which is the conventional hydrocarbons, fossil carbon fuels, and [going] off into what we need to do to save ourselves.

—Dennis Bushnell, chief scientist, NASA Langley Research Center, in an interview with the author, February 2, 2008

By 2017, world demand for oil, which peaked in 2013 at ninety-five million barrels per day and US$375 a barrel, had dropped back to seventy-six million barrels per day, and the price of oil went into free fall. Although global production was also falling slowly (everyone agreed by now that "peak oil" had been around 2012), demand was falling faster, and even the mothballing or outright abandonment of some high-cost operations like the Alberta tar sands and the tail-end of the North Sea fields did not bring supply and demand back into balance. By 2019, oil was selling for only US$30 a barrel (US$18 per barrel in 2006 dollars), and there was not a single oil-exporting country in the world whose budget was not in ruins.

The first revolutions came in the countries with big populations and relatively low per-capita revenue from oil income: Nigeria in 2017 and Iran in 2019. They were exceptionally bloody revolutions, in which people who were believed to have grown rich from stolen oil revenues during the good years were hunted down by mobs and killed: at least a million died in Nigeria, and in Iran, the radical neo-Marxists who seized power wiped out virtually the entire clerical class, including their extended families. Other oil exporters, which had managed to build up very substantial reserves during the fat years, were better able to roll with the punches, but the resident foreign population of the Gulf states began to move out very fast after 2017, and after the Iranian revolution a good many of the citizens followed them.

This proved to be a prescient move, as the Israeli-Iranian War of 2021 put paid to the erstwhile prosperity of those states. Israel targeted only Iran in its original "pre-emptive strike" and avoided using nuclear weapons on cities, but after Haifa was nuked the Israelis flattened Iran—and used mini-nukes and precision conventional strikes to take out key oil terminals and pumping stations all the way down the Gulf. But the rest of the world's loss of interest in Middle Eastern oil meant that there was little chance of the war spreading outside the region. The loss of practically all oil exports from the Gulf only brought the price of oil back up to US$100 a barrel, and only for a few years.

Getting non-oil emissions down globally was a much slower process. Most older industrialized countries were producing at least 20 percent of their electricity from renewables by 2020, and had saved around another 20 percent by stringent conservation measures, but this barely compensated for the added emissions caused by continuing rapid economic growth in Asia, home to half the world's population. Carbon dioxide emissions, which had been growing at more than 3 percent a year as late as 2010, were actually falling by 2 percent a year by 2020, but that was not enough to meet the "Zero-2030" target, or indeed a Zero-2050 target. The long-delayed conclusion of an international deal to protect the world's remaining tropical forests promised faster cuts in carbon dioxide emissions in future, since the ongoing destruction of those forests had once accounted for about one-fifth of annual human greenhouse-gas emissions, but it would have been a lot more help ten years earlier: by 2020, almost all the tropical forests outside the Amazon and Congo basins were already gone.

And the climate-related disasters kept coming: the early 2020s saw Europe's first-ever hurricane (quickly followed by two more); the virtual collapse of agriculture in much of Central America and southern Mexico as the drought became semi-permanent; and immense loss both of life and of land on the cyclone-battered Bangladesh coast. The mini-states of Tuvalu and Kiribati in the southwestern Pacific, both low-lying island groups, were officially evacuated because of rising sea levels, with most of the inhabitants moving to Australia or New Zealand (although many islanders refused to leave). Most disturbingly, the now-regular summer disappearance of Arctic sea ice raised average summer temperatures in northern Canada, Alaska and Siberia by five degrees Celsius above the long-term average for the period 1978–2006, leading to rapid melting of permafrost and massive releases of both methane and carbon dioxide into the atmosphere.

The cumulative effect was to induce a sense of despair in many people who had been enthusiastic in their support for mitigation measures ten years before: nothing seemed able to stop the warming. It was irrational, but it was entirely human, and popular support began to drain away from political parties and personalities who had spent the past ten years urging everybody to make sacrifices and work hard to contain this problem. This was compounded, in the old-rich industrialized countries that had done most by their emissions to cause the problem, but were so far suffering least from its consequences, by what can only be called "compassion fatigue." Everyone was horrified when almost a million Bangladeshis died in the great cyclone of 2022, and the country was overwhelmed with aid and sympathy from abroad. When half a million Bangladeshis drowned in another cyclone the following year, there were still expressions of sympathy, but a lot less aid. And when Cyclone Anwar struck in 2025 and killed at least two million, a lot of people in the rich world simply wondered why the stupid Bangladeshis insisted on living in such a dangerous place, as if there were lots of unoccupied lands elsewhere for them to move to. Shameful behaviour, of course, but also quite human.

Global cooperation on curbing emissions had been very hard to achieve even in more optimistic times. That was why China and India had still not committed to a formal target for cutting emissions (though they were working very hard on it in practice), and why the tropical-forest treaty had taken over a dozen years to complete. Now, as hopes for an early return to climate normality faded, even the limited global cooperation that had been achieved was starting to break down. Drought-stricken regions were not getting enough food aid (which was still theoretically available, despite the global food crisis). Victims of climate-related disasters were not getting prompt and adequate aid. The rich countries were abandoning the poorer countries and pulling up their drawbridges, even though most people in the wealthy countries would still admit (if pushed up against a wall and threatened with a knife) that only global cooperation could possibly stop the rise in emissions before the climate hit various tipping points and went into runaway warming.

> If we move forward into perhaps 2020, where we have to do the real global deal . . . the issue will become more and more about how we manage the fact that climate change is going to get worse and worse while we're doing more and more on mitigation. No matter how fast we decrease carbon, the climate gets worse for another forty years. We've never tried to do politics like that before, and this will take some real far-sighted statesmen and women to realize that, if you're going to keep the deal going while everybody tightens their belts but no one sees returns, you're going to have to deal with the immediate consequences like food security, and show that there is some innate fairness.
>
> Climate change is unique. It raises enormous numbers of hard-security problems, and it has no hard-security solutions. In fact, the only solution it has is cooperation. So it's the realpolitik of the generation born in the sixties. This is as hard as cooperation gets. Cooperation is difficult, it's emotionally draining, it's boring—just look at Europe—but it's the only way out. It's easy to be nice and immature, to sit there and say "We're going to defend our borders and be independent" and blah, blah, blah. Much simpler politically and much simpler emotionally, for most of our actors. It takes real maturity to do cooperation. We need to realize that our hard-security future will only be preserved by doing this boring, drawn-out, messy [negotiation]. Don't do Henry V before you go into the conference chamber. And we need to have a cadre or generation of leaders who get that. That's the hard yards.

That's the World War Two of their generation. It's not
glamorous, but it's incredibly important.

We spent hundreds of years learning about how to
manage the Industrial Revolution. We've only got one
shot to handle climate change, because if we miss the
target, the climate takes over itself, and all we've got left
is adaptation, which will be brutal and ugly.

—Nick Mabey, director, E3G,
in an interview with the author, March 14, 2008

It was the Bangladeshis who broke the deadlock. They
threatened to upload a million tonnes of powdered sulphates
into the stratosphere—in order to cut incoming sunlight and
drop the global temperature unilaterally—if there were not
swift global agreement on doing it by less noxious means.
There was no doubt that they were serious about their threat,
that they had the technical ability to carry it out—and that any
attempt by the rich countries to stop them by a pre-emptive
attack would divide the world into hostile North and South
blocs. Indonesia offered to mediate, and emergency negotiations in Jakarta led to a deal by which the old-rich countries
would pay for a variety of urgent "geo-engineering" measures
to stop the rise in average global temperature, and subsequently to bring it back down towards the old normal. Cloud
spraying, oceanic iron fertilization and a somewhat more controllable version of Bangladesh's sulphates-in-the-stratosphere
solution would be implemented at once. Long-term development work would start on building space-based mirrors to control the Earth's temperature.

Meanwhile, all the measures for cutting greenhouse-gas
emissions that were currently in force would be maintained
and, if possible, accelerated, with the ultimate goal of reducing
the amount of carbon dioxide in the atmosphere to something
resembling the pre-industrial level. The difference was that in
the meantime the world's human population (and its ice caps

and wildlife and forests and oceans) would be protected from the worst impacts of a major further rise in global temperature.

It was a near-run thing, but the Jakarta Treaty of 2026 was a turning point. Cyclones continued to batter Bangladesh with unprecedented ferocity for another ten years, before the gradual fall in sea-surface temperature removed the extra energy from the weather systems, and one-third of the Amazonian rainforest burned in 2029 (the smoke from the fires rose into the stratosphere, turning 2030 into the "Year without a Summer"), but gradually the number and scale of the disasters diminished. By 2050, the world's major economies were effectively carbon-neutral, although it took much longer to get the carbon dioxide concentration in the atmosphere back down where it belonged, and the world's oceans were still suffering from serious acidification. It was 2075 before the carbon dioxide concentration fell back to the 2008 level of 387 parts per million, and it was estimated that it would not reach the interim target of 350 parts per million until almost 2100.

Then there would be a long, complicated argument about what the long-term correct setting for the carbon dioxide concentration should really be. Does the human race want to extend the endless summer of the current interglacial indefinitely, and stop the glaciers from ever marching south again across the Northern Hemisphere? Do we want to geo-engineer our way out of the ice age that we were born in? And, no doubt, there will be a faction of people who argue that we should not interfere further with the climate system. We should turn the dial back to 280 parts per million, where it was at the start of the Industrial Revolution, and let nature take its course, even if that means that the glaciers start moving again in a few thousand years. But that is an argument for the next generation, or the one after that. This generation has done its job. It has saved civilization.

Editors' Note: This curious document was recovered from a "computer hard drive" discovered by gill-divers exploring ruined houses in the shallows of North London Bay, in a district once known as Camden Town. The document appears to have been written at least fifty to sixty years before the Inundation. Our knowledge of early 21st-century history is inevitably patchy, given the level of destruction that occurred in the latter part of the century, but it is unlikely that the events described in this document ever actually occurred. If such large changes had really been made early in the twenty-first century, Britain would still be a single island today. It is best treated as a fantasy, or a pious hope.

CHAPTER FIVE
. . . But Probably Not in Time

Technically the climate problem is very soluble, and I don't mean by geo-engineering; I mean just by stopping emissions. There are lots of ways to decouple human energy use from emissions—by wind power, solar power, nuclear power—lots and lots of things we can do, and more will be invented. And the costs of doing all that are, I think, quite reasonable: they're a couple of percent of GDP [gross domestic product] . . . Now, personally, I'm totally sold on spending that because I think we'll get a great deal in terms of environmental protection and reducing the risk to my kids and so on. So why am I worried that we might not succeed?

I'm worried that we might not succeed because of reasons that are essentially game-theoretic. The sad fact is that the optimal strategy for each country is to get other countries to cut their emissions while each country does nothing. The sad fact is that if you spend a lot of money to cut emissions in your country, you're distributing the benefits of that cutting all over the world, but all the costs of the cutting are in your country. This is the way an economist would think about it.

Another sad fact is that one of the things that people talk about a lot now, more and more publicly, is the difference between mitigation—cutting emissions—and adaptation: dealing with the climate change. But when you think about this from the perspective of a national government, if you spend money on adaptation in your country, you know the money will be spent in your country, and the benefits will be there. If you spend money on mitigation, those benefits are being spread around the world. For those reasons, I think it's very, very hard to crack this nut.

Humanity has managed to crack some of these global pollutants like carbon dioxide before, most notably the chlorofluorocarbon, ozone-hole problem, but that was a lot cheaper than this problem, so I think there's room to worry . . . In the last fifteen years, we've signed this wonderful treaty, the Framework Convention on Climate Change, and Kyoto, and we've had an enormous amount of grandstanding talk about this topic, and meanwhile, the rate of growth of carbon dioxide emissions has gone from a little over 1 percent to somewhat over 3 percent, and we are now on an incredibly fast growth path, heading up towards doubling or tripling carbon dioxide concentrations in the atmosphere in the lifetime of a single human being—my kids. There's a lot of inertia in the world's energy system, and it's pretty hard to stop.

Another reason why this is so hard is that this is fundamentally about taking money from today's generation and giving it to the next generation. Think about the difference between this and cutting air pollution in the city. When you cut air pollution in the city, you've got to make a lot of people pay a lot of money: companies have to pay to put scrubbers in their plants, everybody's car is going to cost more. And there's benefits to that, of course, but the neat thing is that the benefits accrue to the same generation that spent the money. You put in all those scrubbers, and within a couple of years, the air gets cleaner and your kids are healthier.

The climate problem has such a long time dimension to it that if we work very, very hard [at] cutting emissions for the next thirty years, the generation that spends that money for thirty years will see no benefit at all. There'll be a big benefit later in the century, because there will be much less carbon in the atmosphere than there would have been if they hadn't cut the emissions, but there's no immediate benefit that you get from cutting emissions now. The benefit is spread into the future, and that's the second great reason why this is so hard to do. People talk a lot about spending money for future generations, but typically they don't do it very much.

—David Keith, Canada research chair in energy
and the environment, University of Calgary,
in an interview with the author, May 2, 2008

David Keith's research is all about coming up with fallback technologies to get us through the crisis if conventional means of reducing our greenhouse-gas emissions don't get them under control in time, so a cynic might say that he has an interest in seeing them fail. I don't believe that for a moment. He's simply pointing out something that's as plain as the nose on your face. This problem may be technologically easy to solve, but because human beings are what they are, it is very difficult to deal with politically. Genuine altruism is rare, deferred gratification is very hard to practice (especially when it is deferred past the point of an individual's death), and we all know about the "tragedy of the commons." What Winston Churchill said about Americans actually applies to the whole human race: you can count on us to do the right thing in the end, but only after we have exhausted all the other alternatives.

This applies in particular to some of the technological "solutions" that we have fallen for. We have already discussed the biofuels scam, but for sheer delusionary persistence, nothing compares with the ever-receding mirage that travels under the name of "Carbon Capture and Sequestration." If CCS could be made to work at an economically feasible cost, then all our electricity-generation problems would be solved at a stroke: we could just go on burning all the coal we want, and bury the resulting carbon dioxide deep underground forever. We could still be a coal-fired civilization in the twenty-third century, if no cheaper way of generating energy had presented itself. But there is a large "If" buried in the middle of this paragraph.

It is the "magic bullet" aspect of CCS that persuades so many people to believe in it. For example, George Monbiot, having done his brutally realistic analysis of how Britain might cut its greenhouse-gas emissions by 90 percent by 2030, finds in the end that he still needs to run quite a few gas- or coal-fired power stations, even after all the wind farms have been built and all the conservation measures have been taken. He could choose to replace most of them with more nuclear-power stations but, instead, he gambles that carbon capture and sequestration will be a mature and commercially viable technology in time to allow Britain to continue to run significant numbers of coal- and gas-fired power stations into the third decade of this century and still meet his target. I cannot prove that he is wrong, of course, but I would observe that carbon capture and sequestration, like Iranian nuclear weapons, is always five to ten years in the future.

This is curious, because every element required for CCS to succeed has already been demonstrated to work experimentally: separating and capturing carbon dioxide and other greenhouse gases from the exhaust flues of coal- and gas-fired power plants; compressing the carbon dioxide gas and moving it by pipeline to some suitable location for sequestration; and pumping it down into some geological formation—an old oil or gas field, a coal seam that cannot be mined, or a saline aquifer (an underground pocket of salt water)—that will hold it underground indefinitely. There are even a couple of places where something quite similar to CCS is done on an industrial scale, and Monbiot, like everybody else who has been driven to believe in the viability of CCS, names the usual suspects: the operation by the Norwegian oil company, Statoil, that scrubs a million tonnes of carbon dioxide a year out of the natural gas it extracts from the Sleipner field under the North Sea and pumps it into a saline aquifer under the seabed; a rather similar BP operation in Algeria; and EnCana's operation in Saskatchewan, where carbon dioxide is piped three hundred kilometres under pressure from a coal gasification plant in North Dakota and pumped into an oil reservoir in order to flush out the remaining oil.

These three places are the only CCS projects on the entire planet, after all these years in which CCS has featured so prominently in the climate debates. The operation that most closely resembles what would be needed to sequester the carbon dioxide emissions of power plants is EnCana's, so I asked them what conclusions could be drawn from their operation. I spoke to Gerry Protti, executive vice president, corporate relations, and president, offshore and international, at EnCana Corporation, on May 1, 2008:

> A thousand megawatt coal-fired facility would produce about eight megatonnes per year of carbon dioxide. Now, how much of that you capture depends on the technology utilized, but you cannot capture all of it that's going up the flue stack. Because it's so diluted, it's very expensive to capture. It doesn't have the purity of the process stream that a gasification plant would have, or a process stream from a fertilizer or petrochemical plant.
>
> The industrial processes that have a pure stream of carbon dioxide are obviously the lowest cost in terms of capture. Once you've captured it, the cost associated with transportation and injection is relatively small; it's about $1 or $2 a tonne to transport and about the same amount to inject. The question is, if you're spending a lot of money to capture from a source that has low purity, like the flue gas stream out of a coal-fired plant, do you have a place to put it where somebody's willing to pay for it?

Right now the only place is in an enhanced oil-recovery project, so in the first instance, carbon capture and sequestration will have to be anchored by enhanced oil recovery projects. Otherwise society is going to be paying a tremendous cost to take this relatively low-purity carbon dioxide and just inject it into the ground and treat it as a waste product. If you're trying to capture it out of the coal-fired facility that we described, you're into the hundreds of millions, the billions, very quickly if you're doing it over several projects. Society will have to determine whether or not that's the way it wants to spend its finite resources.

The problem that will not go away is the energy cost of extracting the carbon dioxide and then compressing it in order to transport and inject it. At the moment, using the standard amine method, 29 percent of a coal- or gas-fired power plant's output would have to be devoted to the capture and compression of the carbon dioxide that is produced in the combustion process. (In the trade, it's called "parasitic load.") What that means is that the coal- and gas-fired sector of the electricity generating system would have to be almost a third bigger to deliver the same power to consumers if full carbon capture and sequestration were required. Gas-fired power stations produce only about half the carbon dioxide emitted by coal-fired stations of the same rating, but the CCS costs are roughly the same.

A new chilled ammonia system, owned by Alstom Power Systems, allegedly cuts the cost of separating the carbon dioxide dramatically and is scheduled to be tested in a coal-fired power plant in northern Alberta. However, almost every other CCS pilot project that involved a full-scale power plant has been cancelled or put on indefinite hold—and nothing will remove the cost of compressing and cooling the captured gas. Compression heats the gas, and it must be cooled each time before proceeding to the next stage of compression in order to get it up to the very high pressure (two-thousand-plus pounds per square inch) needed to inject it into the ground. There is no magic bullet. When the technology is mature, in ten or fifteen years, CCS may work even as a retrofit on existing power plants, but it will only work at a very high cost, and it will not be widely available in time to get Monbiot anywhere near his goal of 90 percent cuts in British emissions by 2030.

I have no secret plan that is better than Monbiot's. And, to be fair, he does offer options other than the Hobson's choice between unpalatable nuclear power and implausible CCS. In a world in which renewable resources can be exploited cost-effectively, solar power may well be exported from the southern shores of the Mediterranean to the less sunny nations of northern Europe. A European Union-wide deal with North African countries to finance huge solar power farms in the deserts and move the electricity to Europe via undersea cables, already in development, may be the first step towards a power grid that will tolerate a much higher percentage of renewable energy than it now can. The wind may fluctuate wildly at times in the seas around Britain, but the sun shines pretty reliably on the Sahara, so, if it were all tied together in a "super smart" grid covering all of Europe and North Africa, the proportion of renewables in the system could be much higher. The secret to moving electricity over such long distances without huge power losses is to switch from alternating current to high-voltage direct current cables for long-distance transmission—and indeed, there are people in the European Union who are working right now on creating the economic and legal framework for exactly such a deal.

The trans-Mediterranean grid is only the first step towards a pan-European grid. This obviously can be duplicated anywhere else in the world: in China, in India, in Brazil there are already [high-voltage DC] lines. You ask, what is the advantage? With high-voltage DC lines you can transport electricity from point to point with minimum loss. As of today the technology is available, it is already deployed, and it is economically reasonable.

The problem with renewables is the fluctuation of the load. Once you have these new lines, you overcome this fluctuation problem, and therefore you can really put in the grid renewable electricity from many different sources. Europe has committed to [having] more or less 60 or 70 percent of its electricity produced by renewable energy sources by 2020. It has also committed to [having] 80 percent reduction of emissions by 2050. For the majority of the countries in Europe, this is a huge challenge. The super-smart grid would make it easier for many countries to meet the 2020 target, and would be absolutely essential for countries like Italy, but it would also set up the base for realizing the 2050 target. The grid which extends to northern Africa has the possibility of transporting electricity from renewable [solar] energy sources of immense potential.

—Antonella Battaglini, scientist and researcher,
Potsdam Institute for Climate Impact Research,
in an interview with the author, March 27, 2008

But I'm not really trying to second-guess George Monbiot. The point is rather that CCS is a perfect illustration of the fact that nobody has a master plan for getting through this at the moment. The grand global visions of how to fight climate change, like Lester Brown's, come with few technological details and no political strategy. The technologists (or at least the Americans among them) believe that the market will get it right in the end, even if it does seem to be taking its own sweet time about it. And the blue-sky people, like Dennis Bushnell, argue, probably rightly, that the safest way through is a decision to dump the existing energy technologies as fast as possible and go flat out for radically new technologies. They are all correct in believing that global warming is an eminently soluble problem, but they are probably optimistic in their belief that it can be done quickly enough to avoid much damage, and many unnecessary deaths. The politics is what slows it down and screws it up, but no big thing can be done in the human world without a great deal of politics.

So we are back to the slow, grim grind of international negotiations to get some kind of agreement on cutting emissions, and the brutal truth is that it will be very hard to get new commitments to big cuts in emissions in the forthcoming round of negotiations for a post-Kyoto deal, precisely because the situation now seems so much more serious and urgent. Back in 1997, it was relatively easy for a national leader to sign up for 5–10 percent cuts in emissions over the ensuing fifteen years; almost certainly he or she would not even be in office any more by the time the obligation had to be met. It is much harder to sign up for, say, 40 percent cuts in ten years, which would be more commensurate with the scale of the threat that now confronts us. Moreover, the question of what to do about the very rapidly developing countries that were exempted from any constraints on their emissions under the 1997 Protocol has grown into a monster that may devour the treaty.

In the year 2008, Chinese greenhouse-gas emissions probably overtook those of the United States. China's per-capita emissions are still much smaller than those of the United States, of course, since its population is about four times as big, but it becomes harder and harder to defend the notion that China should remain exempt from any commitment, however fuzzy, to control its emissions. The same is true for India, Brazil, and other rapidly industrializing ex-Third World countries. But it remains true (a) that they are still relatively poor countries, and the rapid growth in their emissions is mainly due to the fact that they are becoming less poor; (b) that the old rich countries are responsible for almost all of the emissions that have created the current problem; and (c) that China will accept no commitment unless the United States does.

Those three facts may well spell failure for the post-Kyoto process. The rapidly industrializing countries have a large and entirely justifiable chip on their shoulders, because the extra hundred parts per million of carbon dioxide in the atmosphere that have brought us to the brink of runaway climate change were put there over the past hundred-odd years by the old industrialized countries. They are being asked to contribute to the solution to somebody else's mess: if China and India had been the first countries in the world to industrialize, their emissions would not become a problem for a hundred years. So any deal that requires them to curb their emissions at this stage in their development will have to be a highly asymmetrical bargain, in which the older industrial countries make far deeper cuts in their own emissions, and also heavily subsidize the cost of decarbonizing the economies of the developing countries.

The great hope for the new Kyoto round is that, after 2009, there will be a new U.S. administration that plays a positive role in the process rather than deliberately trying to sabotage it, but it remains to be seen whether that will make a big enough difference to change the likely outcome. Asymmetrical deals are particularly hard to negotiate, although the basic principle that must underlie this one has been clear for some time. It is that each person on the planet has an equal right to pollute the atmosphere, including an equal right to emit carbon dioxide. In some circles, it is seen as a radical notion even now. Twenty years ago, when four men meeting in musician Aubrey Meyer's house in Walthamstow, North London, first formulated the principle, it was revolutionary. I spoke to Meyer, founder of the Global Commons Institute, about this on January 16, 2008:

> In 1988, I was looking for the subject of a musical, and I came across this report on the murder of Chico Mendes [the Brazilian campaigner for the protection of the Amazonian rainforest], and my initial reaction was "Oh,

great! This is a human interest story, drama, Latin American music, this is perfect!" So I couldn't find out much about him, but I found out about the issues, and within a matter of two weeks, I was crawling around the floor of the flat in tears, thinking "Jesus, we're completely stuffed. This is deadly serious."

So I took a mad decision and just thought: "Fiddling while the planet burns is a waste of time. You've got to try and do something." . . . Within a year we'd set up the Global Commons Institute. The formula addressing climate change and global collapse was very simple: equity and survival. You couldn't untie that. Not even quantum physicists could untie it. It was obvious. You had to make a deal counting everyone in as equals.

The fundamental principle of "contraction and convergence" is equal rights to emissions under the overall limits that save us. To stabilize concentrations in the atmosphere you, by definition, have to have a deep contraction in emissions. It's like turning off the tap in a bath. You have to turn the tap right off to avoid it overflowing. It's non-negotiable. You've got a little wiggleroom with the plug, but the plug, in this case—the natural carbon sinks—is blocking up, so we've got to get on with it.

And then the issue is: inside that bath, whatever the very unequal shares have been historically till now, the only conceivable way to sort this is on the basis of an equal share on everybody's account to use the atmosphere, which is a common resource for everybody. "Convergence" to equality is a way of just softening that, because to try and do it overnight would be a bigger wrench in the system than any of us could possibly organize. But you can programme the system to go to there—it's not the best option, it's the least worst. What's the alternative?

There is no alternative, because human beings care intensely about fairness—more, sometimes, than about their so-called real interests. It is simply inconceivable that Chinese and Brazilians and Indians and South Africans, as their economies develop and their emissions increase, will accept the notion that the old industrialized countries can permanently retain a bigger per-capita right to emit greenhouse gases. There must be convergence towards equal shares for all, or there will be no deal.

The idea behind "Contraction [of emissions] and Convergence [of rights to emit]" is now mainstream, and like all successful ideas it now has many would-be fathers, but it was Aubrey Meyer and his Global Commons Institute who took it to market and sold it. This notion that equity demands a global transfer of resources from those who pollute more to those who pollute less, on condition that those resources are used to minimize the growth in emissions as those poorer countries grow economically, is now at the basis of almost all serious negotiations between the countries of the North and the South on Kyoto-related topics, even if the Northern side is still not very comfortable with it.

But it *is* an asymmetrical deal, and the United States, which has been largely absent from these conversations for the past eight years, will have great difficulty in accepting it even under new management. Moreover, no deal will be worth the paper it's written on if it does not tie the cuts that are negotiated to a clearly defined target of how much carbon dioxide we can tolerate in the atmosphere. The Kyoto tradition of seeking proportional cuts to existing emissions, with no reference to any scientifically based target, is what prompted Aubrey Meyer, in my interview with him, to say, rather frankly, that "If the Kyoto Protocol is the best that the evolutionary process can provide as an example of the survival of the fittest, the very clear deduction is that we're not fit to survive, and we're not going to. We have so lost our way."

If we must never exceed two degrees Celsius warmer than the pre-industrial average global temperature, and that equates to an upper limit of 450 parts per million of carbon dioxide in the atmosphere, then any post-Kyoto deal that does not promise to achieve that goal in a timely fashion is worse than no deal at all. It would just lull people into a false sense of confidence that "something is being done." Much better would be a spectacular failure for everyone to agree, because that at least would make the situation plain. And spectacular failure may well be what comes out of the negotiations on a post-2012 accord that are supposed to conclude in Copenhagen at the end of 2009.

In practice, success in the post-Kyoto negotiations depends on there being a prior deal between the major industrialized countries and China, the biggest of the rapidly developing countries. That will be rather difficult to arrange, as Nick Mabey, who was head of sustainable development at the British Foreign Office and a senior adviser in the Prime Minister's Strategy Unit before founding E3G, explained to me when we spoke on April 14, 2008:

> To get the first concrete step to the next global deal on climate change, we need to make a deal with China. China does certain things that are monitored under an international agreement, in return for technology, and a little bit of money, and a certain amount of trade, and we will do a lot in return. At the moment, there is no public agreement that we should give China *anything* to deal with climate change, neither in the U.S., nor increasingly in the EU. But if China doesn't get anything, China does nothing—and nor should it, it's much poorer than we are.
>
> [We have to] overcome what is essentially a politics of fear, which is all about textiles and, you know, future Chinese missiles in 2050—it's nothing to do with climate change, it comes out of a wellspring of fear about China's emergence—and legitimate dislike of what they do in Tibet and issues around human rights.

But if that's going to get in the way of us doing a deal with China, then Tibet's gone, too. So whose human rights are you preserving?

That's a pretty tough discussion to have with a lot of people—the public, and security decision-makers—in the EU and the U.S. But we're going to have to do it, because if we don't, we'll fudge it. We'll write a deal on paper [in Copenhagen], but it won't be a deal in substance, and then we'll wonder why things are getting worse and worse over the next decade. If we can't do the China question right in the next two years, then we might as well just go home, because we'll have missed all our deadlines.

Mabey has spent a lot of time in the diplomatic and policy trenches, and he is quite right to emphasize how hard it will be to make that deal with China and sell it to the folks back home. He is also right to stress the urgency of the situation: the amount of climate change that the world undergoes in the next half-century will largely be determined by decisions that are made in the next five to ten years. But the game is not lost yet, and even insiders, such as John Holdren, the director of Woods Hole Research Centre, can muster up some optimism when it's required. When I spoke to him in February 2008, he was confident that there would be a global response:

I do think that the United States will make the transition from laggard to leader in the climate issue in the next couple of years. Whether we get a Republican or a Democrat in the White House, it's going to happen either way. I think when the United States finally does that, when the United States imposes a mandatory economy-wide set of restraints on greenhouse-gas emissions—it'll probably do that in the form of emissions limits implemented through tradeable permits, the so-called cap and trade approach—and when at the same time it starts to boost very substantially the country's investments in research and development on technologies that can help us meet those caps in an affordable way, I think a lot of other countries are going to follow.

One of the questions that is often asked is: What are you going to do about China and India? They're only interested in development. They're not going to buy into this. Their emissions are going to keep growing . . . I don't believe that for a minute. I spend a lot of time in both China and India. I run research projects in collaboration with government organizations, think tanks, universities in China and India on climate change and what to do about it. And what I can tell you is that the Chinese and the Indians are not less knowledgeable and not less worried about this problem than we are in the United States or Canada or Europe. They are waiting for us to lead, in part because we, in the industrialized world, caused most of it up until now. But they understand that climate change is already harming them.

The Chinese have figured out that the East Asian monsoon has been changing for thirty years in a manner predicted by the climate models . . . and that has caused increased flooding in the south of China and increased drought in the north. Chinese climate models show this. Chinese leaders know it. You go and sit privately with the political leaders of China and they will quote to you the results of their own Chinese climate scientists' studies showing that China is being seriously harmed today by climate change. It's the Chinese themselves who figured out that the great glaciers that feed their rivers are disappearing at the rate of 7 percent per year. That's a halving time of a decade . . . The Chinese and the Indians, in my view, are going to sign on to a global approach to reducing greenhouse-gas emissions within three to five years of the United States making the transition from laggard to leader. They're waiting for us to do it, but they're going to join.

The game is not lost. There will almost certainly be some serious steps taken to curb emissions in the next few years, both within individual countries and internationally. But it is almost equally certain that they will not be enough to avoid further climate change. It is very hard to believe that either the post-Kyoto accord or the domestic restraints on emissions imposed by the next U.S. administration will involve 80 percent cuts by 2030. With luck, they might aim to do half of that—in which case, we can look forward to only about half of the calamities that a complete failure to respond to the problem would entail. And then, in due course, some of those calamities will motivate us to make deeper cuts—although, coming later, they will be of less use. And so on. It is probably going to be a long, miserable experience, with an uncertain outcome.

CHINA, 2042

Today, most of the work on the security implications of climate change assumes that they will happen elsewhere . . . but actually there will be very real security consequences here, in the U.K., and in other developed nations . . .

If the environment is changing very rapidly and government-policy responses are not keeping up with it, then people may lose confidence in the government's ability to protect them . . . There'll be some people who believe the government aren't going far enough in their changes . . . but there will also undoubtedly be a segment of the population who believe that the government are going too far . . .

If we look at some of the more extreme social and environmental groups that exist currently in Britain — I'm talking primarily about animal rights activists — then, as the changes begin to happen, these people may become more and more radicalized around environmental issues. We may see a rise in non-violent direct action, and, potentially, some of these groups turning to terrorist-type tactics in the extreme. We've seen this sort of thing in America, where, perhaps unfairly, many groups have been labelled eco-terrorists. It's something I know the FBI are very concerned about.

—Chris Abbott, Oxford Research Group,
in an interview with the author, March 26, 2008

THE DEEP ECOLOGISTS AND THE ZERO FOOTPRINTERS are understandably furious when anyone suggests that the Toba Winter was a consequence of their policies, because nothing could have been further from their goals, but it is hard to deny that there was a causal connection, however unintended. They are even angrier when their name is mentioned in the same breath as the eco-terrorists, whom they rightly denounce, but in political terms there is a link there, too.

A decade of large-scale terrorism in the 2030s has pushed many Europeans and North Americans into a crude and paranoid world view that lumps all "terrorists" together as an irrational, even inexplicable, phenomenon that is simply "evil." The same phenomenon was observed in the earlier panic about terrorism in the first decade of the century, particularly in the United States, and it was not conducive to clear thinking and sound policy then, either. But there is an important difference between eco-terrorist attacks and those carried out by terrorists from the formerly oil-rich countries of the Middle East.

The bombing of the Tunisian terminal for the cable carrying power from the Saharan solar arrays to Europe in 2031, for example, was certainly terrorism, but there was virtually no ideology involved. Interrogation of the surviving attackers, all from the Gulf countries that had been devastated economically by the collapse in the demand for oil, suggested that simple rage at the evaporation of the prosperous future they had expected for themselves was their primary motive.

The twelve-person team who attacked the Number 2 reactor at Penly in Normandy in 2032, on the other hand, were genuine eco-terrorists. They were native-born French citizens (and one Belgian), well-educated and leading comfortable lives, whose motives were specifically ecological and ideological.

The horrors that ensued when the main plume of radiation from the burning reactor enveloped Paris have made them monsters in popular memory, but the investigation made it clear that they were motivated by profound opposition to many of the measures that had been adopted to fight climate change. Their hostility to the spread of nuclear power as a substitute for fossil fuels was only a symbol of a much broader disaffection. (Ironically, the Penly reactors, which had run continuously for over forty years, were due to be decommissioned only three months after the date of the attack.)

The distinction is important because terrorists coming mainly from what used to be the oil-rich countries of the Middle East are easily profiled, must make their way through heavily fortified borders, and cannot carry advanced weapons. They do not easily find support among Muslim Europeans, who are mostly descended from immigrants from Turkey, North Africa and Pakistan, and so their attacks have mainly been directed not at Europe but at the softer targets of Egypt, Libya and the Maghreb, whose lucrative deal to sell solar power to the European Union excites anger and resentment among the destitute and desperate populations of the formerly oil-rich states.

The eco-terrorists, however, are already inside Europe's borders, and their motives are so far removed from normal politics that they could be anybody. Unusually for a terrorist movement, their activists include almost as many women as men (like the old Baader-Meinhof Gang, which they somewhat resemble in spirit although not in ideology). And as citizens of a technologically advanced culture, they have access to serious weapons: the Penly attackers used infrared cloaking devices, UAVs coming from the opposite direction to confuse the reactor-site defences, and deep-penetrator man-portable missiles with a two-thousand-metre range. Moreover, their sympathizers are everywhere: their weapons must have come from supporters in the armed forces, and they may have had a collaborator inside the Penly site who took the movement detector system off-line about ten minutes before the attack. (Since nobody in the control room survived, this has never been confirmed.)

The eco-terrorists are not simply anti-nuclear extremists. They have their roots in the great debate that split the West in the teens and twenties, between those who wished to retain as much of the old way of life as possible, solving the "climate-change problem" by a combination of conservation measures and various technological fixes, such as carbon capture and sequestration (CCS) and more nuclear power, and those who argued that the old way of life *was* the problem and that far more radical changes were necessary.

It would be unfair to characterize those on the "right" (as they were inevitably called) as stubborn technophiles. Most had switched to an electric car or no car at all, and accepted even radical conservation measures, such as the cuts in air travel, as necessary. They simply argued that the current density of population could not be maintained without very large amounts of energy—so technological fixes in the energy sector were absolutely necessary.

The "left," similarly, were not all anti-technological Luddites. The great majority were people who just wanted to shift the emphasis further towards conservation, who did not want any more coal-fired power stations, even with the promise of CCS, and who hated nuclear energy. They had rational reasons for that—few people on either side of the argument were entirely comfortable with nuclear power—but there was also a strong emotional component dating all the way back to the anti-war movement of the twentieth century. Anti-nuclear power was the one issue that united all the disparate strands of the left, and so it got a lot of play.

> Coal-fired and nuclear and gas-fired central power plants are all dying of an incurable attack of market forces . . . Basically what happened is that central plants cost too much and have too much financial risk to be attractive for private capital . . . That's why, in 2006, [the nuclear industry] worldwide installed less capacity than solar cells added; a tenth as much as wind power added, thirty or forty times less than all forms of micro-power. Just

distributed renewables got US$56 billion of private risk capital, nuclear got zero; it's only bought by central planners. There are no orders in the United States, despite subsidies now roughly equalling or exceeding the total cost of the (nuclear) plants.

—Amory Lovins, cofounder, chairman and
chief scientist, Rocky Mountain Institute,
in an interview with the author, May 5, 2008

The people in the Department of Energy have done everything possible to block decentralized power. All those civil servants know that when they retire there'll be a job for them on the board of British Nuclear Fuels. They know Greenpeace isn't going to give them £40,000 a year for doing two days a week on the bloody board, so they're covering their arse for their future, and advising ministers accordingly. [If Greenpeace could offer them the same kind of money, they'd] most probably change their bloody advice.

—Ken Livingstone, former mayor of London, 2000–08,
interview in *The Observer*, March 23, 2008

Neither side really lost the long argument that divided both Europe and North America from about 2010 onwards, in the sense that both conservation and techno-fixes played a large part in the emissions-cutting strategy of every major Western country. But the left felt that it had lost, because almost all the larger countries launched major programmes for building new nuclear reactors during the 2010–2015 period—perhaps because they had concluded that coal plants plus CCS were going to be even more expensive than nuclear power, or maybe for other reasons. ("Other reasons" was the answer strongly favoured on the left, which was greatly embittered as a result.) By the early 2030s, nuclear power plants had sprung up almost everywhere that there was adequate water for cooling the reactors—so while only a tiny handful of extremists ever resorted to violence, they enjoyed a certain sympathy from much larger numbers of people on the left.

That sympathy temporarily evaporated after the Penly disaster, but it revived as the evidence accumulated that the West's strategy on climate change was not working. Oil consumption was down to a fifth of its peak in the early twenty-first century, a large number of coal-fired power stations had been replaced by nuclear plants, a lot of people had made a lot of money out of fighting climate change—but carbon dioxide was already at 480 parts per million by 2035, with much more to come, and killer heat waves were striking Europe and the American Midwest and Southwest several times each summer. Methane was bubbling up out of the permafrost, the oceans' ability to absorb carbon dioxide was down by 70 percent from the 2005 figure, and the Greenland glaciers were sliding into the sea so fast that you could almost see them move.

Those who had argued many years before that the West should aim for even deeper cuts in its emissions in order to draw the emerging industrial countries into a global agreement had been proved right—and they were mostly on the left. On the other hand, the milk had been spilt long ago, and it was too late to go back to that solution now: emergency measures were needed promptly, if a whole series of climate tipping points were not to be passed. So the political struggle in the West moved on to the next stage, with many on the right arguing that it was time to resort to last-ditch "geo-engineering" measures to cut down the amount of solar radiation reaching the Earth's surface and so keep the temperature down, and many on the left reacting with horrified rejection.

Some form of geo-engineering intervention to reduce global temperature by a degree or so was already being advocated by countries of the "Majority World" (most of Asia, Africa, and Latin America), not as a permanent solution but at a stopgap measure to prevent the climate from passing some critical tipping points while the human race struggled to get its emissions down. It was the countries of the former Third World that took the lead in this because, being nearer to the equator, they were already suffering severe losses from storm damage, rising temperatures and collapsing agriculture. But the atmosphere is a single, shared resource, so they could not legitimately interfere directly with the climate system unless the West also agreed—and the West did not.

> Many members of the scientific and technical communities fear that the full effects of various geo-engineering schemes are not fully understood. The failure of the ambitious Biosphere 2 facility is one example of a complex project that was unsuccessful because scientists still have a limited understanding of how earth systems work together; the implications for a failed project of global scale are frightening, and as a result policy-makers are hesitant to embrace various geo-engineering schemes for combating the effects of global warming.
>
> Other criticism comes from those who see geo-engineering projects as reacting to the symptoms of global warming rather than addressing the real causes of climate change. Because geo-engineering is a form of controlling the risks associated with global warming, it leads to a moral hazard problem. The problem is that knowledge that geo-engineering is possible could lead to climate impacts seeming less fearsome, which could in turn lead to a weaker commitment to reducing greenhouse-gas emissions. It could be argued that pursuing geo-engineering solutions sends the message that humans can continue to live out of harmony with the Earth as long as we have enough clever technological solutions to preserve human life. This disregard for the overall health of Earth's ecosystems and natural environments is an affront to proponents of sustainable development.
>
> —*Wikipedia* "Geo-engineering" entry, as of June 3, 2008

Many individuals in the West, perhaps a majority, did agree that it was time for desperate measures. However, the fundamental principle that made a relatively civil dialogue between left and right on climate issues possible in most Western countries was an agreement to avoid raising the question of geo-engineering measures. When a few American, British and Spanish politicians dared to break the taboo in 2037, terrorist micro-bombings of long-haul airliners suddenly resumed after a long hiatus. The politicians quickly shut up again. Urgent though the situation was, Western governments were simply paralyzed, fearing that an open debate on geo-engineering would trigger what amounted to a low-level civil war, so they did nothing.

The Majority World couldn't wait. Having allowed a decent interval for the West to sort out its own internal divisions and begin an international discussion of how much geo-engineering of what kind was needed, and how soon, it acted unilaterally. On March 25, 2039, Indonesia and the Philippines began releasing the high-altitude balloons that would place approximately one megatonne of sulphur per year into the stratosphere, in order to drop the planet's surface temperature by one degree.

It was not a very dramatic sight: five or six giant balloons a day, each carrying a payload of about five tonnes, disappearing rapidly into the sky from each of a hundred different sites scattered across the two island nations. The spread of latitudes both north and south of the equator ensured that the sulphur dioxide was released into the tropical upward branch of the stratospheric circulation in both hemispheres. Within months, it would be distributed throughout the stratosphere in all parts of the world, now broken down by chemical processes into sub-micrometer-sized sulphate particles that reflected sunlight back into space and cooled the Earth's surface. It was some days before the West even realized the scale of what was going on.

It was China that had bankrolled the operation and supplied a lot of the equipment, of course, although Indonesia, at least, would have been capable of doing it alone: the technology was not at all demanding and the annual cost was only about US$25 billion. Because it was being hit even harder by climate change than the two tropical countries, China had the most urgent need and had taken the lead. It also had the military power to ensure that nothing bad happened to the two smaller nations if the Western or Japanese reaction should take a violent form.

In fact, while Western governments greeted this unilateral Asian action with loud protests, most were secretly glad that the Chinese and their allies had acted decisively and broken the stalemate imposed on the world by Western ideological divisions. The Chinese assured everybody that they were monitoring the effects of injecting such large amounts of sulphur into the stratosphere and would stop the process at once if there were unforeseen adverse effects, after which the atmosphere would return to normal in about a year. Besides, they were only seeking to replicate the cooling effect created by natural volcanic eruptions that boosted comparable amounts of the same substances into the stratosphere. What could go wrong?

Elements of the left in many Western countries demanded that their governments use force to stop the "dangerous experiment," but even in the few countries where the left controlled the government, nobody seriously contemplated war with China. It might be a dreadfully crippled giant after the beating it had taken from the climate in the past decade, but it was still a giant. Moreover, many people on the left were secretly curious to see if this kind of geo-engineering could solve the problem that dominated their lives and threatened their futures.

For a year and a half, the balloons rose every day, and by late 2040, a pronounced cooling effect was clearly evident all around the planet. It wasn't all the way back to the "normal" of the late twentieth century, but it was like being back in the 2020s again. Then Mount Toba erupted.

It was a mere burp compared to the last major eruption of the Toba supervolcano seventy-one thousand years ago, which boosted an estimated 3,000 cubic kilometres of ash into the stratosphere and caused a "volcanic winter" of three to six years' duration. That one had left a caldera one hundred kilometres across in the middle of northern Sumatra, but this time the volcano merely cleared its throat. Only about 550 cubic kilometres of ash ended up in the stratosphere, no more than three times the amount that had been produced by the explosion of Mount Tambora, at the other end of what is now Indonesia, in 1815. But that was enough.

The Tambora explosion in 1815 had caused the "Year Without a Summer," in which crops failed all over the world and hundreds of thousands, perhaps millions of people starved—all this despite the fact that the explosion actually lowered average global temperature by less than one degree Celsius. The Toba eruption in 2040 lowered the temperature in the temperate zones of the planet by almost three degrees Celsius—and that came on top of the one degree of cooling that the Chinese and their allies had just achieved by their geo-engineering work. They immediately stopped their work once Toba erupted, of course, but they couldn't undo the effects at once, so during 2041, the world's surface temperature was on average four degrees Celsius colder that it had been the year before—and fully two degrees colder than it had been in 1990. The harvests failed almost everywhere.

There would have been serious problems with the food supply even if Mount Toba had erupted at this scale in 1990, but as the climate changed during the following half-century, crop varieties had also been changed to ones better suited to the hotter conditions—and of course they didn't respond very well when it suddenly got four degrees Celsius colder. World grain production dropped by 35 percent, and despite a mass slaughter of animal herds that summer in an attempt to free more grain for human consumption, when the northern hemisphere winter arrived there simply wasn't enough grain to provide all nine billion people with their minimum daily caloric intake.

No more than three to four hundred million people died directly from starvation, almost all of them in the poorer countries of the tropics and subtropics. However, the political violence and social breakdown engendered by this extreme emergency affected even the most stable and prosperous countries, while regions containing about a third of the world's population just slid into "failed state" status. Although the average global temperature returned to normal (the new normal, that is) by the following year, at least as many people as had died in the famine would be killed in the following five years from the second-order consequences of the "Toba winter": civil war, mass migration and genocide.

In the West, the calamity of 2041–42 merely deepened the existing ideological split. The left argued that it proved they had been right all along: geo-engineering was a dangerous technology that must never be tried again. The right insisted that it had just been an unfortunate coincidence, and indeed that humanity needed a battery of geo-engineering techniques to cope with what the world threw at it. (There were, of course, bitter disputes about how much of the damage had been caused by the human intervention and how much by the eruption.) But in the aftermath of the disaster, the left had the wind in its sails, and governments throughout the West fell into their hands.

Which brings us to the current impasse. Global warming will reach plus 2.5 degrees Celsius by 2050, and everyone on the right and many on the left believe that we are on the brink of various tipping points if, indeed, we have not already passed them. Even the radical further reductions in emissions mandated by the new governments, aided by the collapse of emissions in some newly failed states, cannot possibly halt the rise in temperature fast enough to stop us short of those thresholds. Yet geo-engineering is profoundly discredited, and several major Western powers have let it be known that any country that attempts to intervene in the climate unilaterally will face an attack with nuclear weapons.

God help us all.

> The problem from a game-theoretic perspective of cutting emissions is that you have to get everybody to comply. With geo-engineering you have the opposite problem: you want to get everybody to *not* jump on the bandwagon.
>
> When we think of geo-engineering now, we tend to think about a huge, rich, technocratic power like the U.S. being the one that does this, but some of these schemes might be within the reach of the richest individuals on the planet; easily within the reach of quite poor national governments. So, in fact, you might find that a national government like, say, Bangladesh, at some point when they are really going underwater, says I'm not really very interested in all you rich countries moralizing. I just want to turn down the heat and do it *now*. And different people might have different views about how to turn down the heat, and whether Bangladesh was allowed to do that, and that could lead to international conflict.
>
> I think there are questions about whether we should start thinking about what the norms of international control are. Whether we need some kind of international treaty process perhaps, and it's better to think about that before we're forced to make decisions, than to think about that in some chaotic way after some country decides they want to do it unilaterally.
>
> —David Keith, Canada research chair in energy
> and the environment, University of Calgary,
> in an interview with the author, May 2, 2008

CHAPTER SIX
Emergency Measures

Recent research has shown that the warming of the Earth by the increasing concentration of CO_2 and other greenhouse gases is partially countered by . . . sulfate particles, which act as cloud condensation nuclei . . . [Human-caused] sulfate particle concentrations thus cool the planet, offsetting an uncertain fraction of the [human-caused] increase in greenhouse gas warming . . . [so cleaning up industrial pollution and eliminating the sulphur dioxide emissions] could lead to a . . . global average surface air temperature increase by 0.8 degrees C per decade on most continents and 4 degrees C in the Arctic. Further studies . . . indicate that global average climate warming during this century may even surpass the highest values in the projected IPCC global-warming range of 1.4–5.8 degrees C.

By far the preferred way to solve the policy-makers' dilemma is to lower the emissions of the greenhouse gases. However, so far, attempts in that direction have been grossly unsuccessful. While stabilization of CO_2 would require a 60–80 percent reduction in current anthropogenic CO_2 emissions, worldwide they actually increased by 2 percent from 2001 to 2002, a trend which will probably not change . . . Therefore, though by far not the best solution, the usefulness of artificially enhancing Earth's albedo [ability to reflect sunlight] might again be explored and debated as a way to defuse

the Catch-22 situation just presented and counteract . . .
growing CO_2 emissions. This can be achieved by burning sulphur or hydrogen sulphide, carried into the stratosphere on balloons and by artillery guns to produce
sulphur dioxide.

—Paul Crutzen, "Albedo Enhancement by Stratospheric
Sulfur Injections: A Contribution to Resolve a Policy
Dilemma?" in *Climatic Change*, August 2006

PAUL CRUTZEN is a Nobel Prize-winning atmospheric
chemist who risked his entire reputation by suggesting, in the
now famous article originally published in *Climatic Change*
in 2006, that it might become desirable or even necessary to
put sulphur dioxide into the atmosphere in order to raise the
planet's albedo and thus to avoid runaway climate change. In
doing so, he quite deliberately re-opened the public debate
on the taboo subject of geo-engineering. Showing considerable tactical skill, he backed into it rather than tackling it
head-on.

For most of the time since the Industrial Revolution, he
pointed out, human beings have been putting excess carbon
dioxide into the atmosphere, which caused warming, but
those same industrial processes that burned fossil fuels also
emitted sulphur dioxide and other pollutants, which reflected
sunlight back into space and caused cooling. This "fortunate
coincidence" meant that the sulphur dioxide cancelled out
some of the warming effect of the carbon dioxide: between 25
and 65 percent of it, according to Crutzen's own estimate in
2003, but possibly even more according to a rival estimate by
T. L. Anderson in the same year. But those days are coming to
an end: new laws in almost all the developed countries are
compelling industries to stop emitting sulphur dioxide, which
the World Health Organization estimated was causing five
hundred thousand premature deaths annually worldwide.

Thus the "global dimming" of the sulphur dioxide pollution, which had a cooling effect, is being lost, while the carbon dioxide continues to pour forth—and that's a big part of the reason why global warming is accelerating.

Crutzen suggested that we might get around this Catch-22 by replacing the lost sulphur dioxide in a way that did not damage people's health. Rather than put it into the lower atmosphere (troposphere) where people would have to breathe it, we could inject it into the upper atmosphere (stratosphere). Moreover, since small particles tend to remain in the stratosphere for a very long time, we would only have to use a tiny fraction—a few percent—of the huge amounts of sulphur dioxide that we used to pour heedlessly into the lower atmosphere (where any given particle of sulphur dioxide remains aloft for only a week or so) in the bad old days. But even though Crutzen carefully couched his proposal in terms of substituting a less harmful intervention in the atmosphere for an extremely harmful one that we had been doing inadvertently for over a century, he would probably have been lynched if his Nobel Prize (for his work on the ozone hole) had not afforded him some protection.

The rage that geo-engineering proposals stirs in some quarters can only be compared to the fury that suggestions for expanding nuclear power arouses (often in the same quarters), and Crutzen got his full share of it. Like all scientists who venture into this minefield, he was careful to present his proposal in the journal *Climatic Change* purely as a last-ditch technique for halting global warming if it were racing out of control: " . . . [A]gain I must stress here that the albedo enhancement scheme should only be deployed when there are proven net advantages and in particular when rapid climate warming is developing, paradoxically, in part due to improvements in worldwide air quality. Importantly, its possibility should not be used to justify inadequate climate policies, but merely to create a possibility to combat potentially drastic climate heating." But nothing he said helped, because in the eyes of others, including many of his fellow scientists, he was creating a "moral hazard."

The phrase comes from the financial world, where it is argued that governments should not bail out investors who are about to lose their shirts on some misconceived project, no matter how great the political pressure, because to do so simply encourages more reckless behaviour by other investors who expect that their wildly speculative investments will be similarly rescued if things go wrong. In the world of climate-change science, the moral hazard is that if people (and governments) believe that there is some magic bullet that can stop climate change through technological wizardry, then they will lose their motivation to address the problem the hard way, through reducing their greenhouse-gas emissions. Geo-engineering must not be discussed in front of the children, because if they know about it they will behave badly. Although a few of the most respected climate scientists take a quite different tack.

> The human burning of fossil fuels is geo-engineering. The suggestions that we encourage re-forestation and the use of bio-char and the storing of carbon in the soil—they're geo-engineering, but they're of a fairly natural order, and they have multiple benefits, so nobody would object to those. There are other, more extreme geo-engineering things that we could do—and I say we should of course do all the other things first—but you may get to a point where you see the ice-sheets are on the verge of collapsing. Then you have to consider these other possibilities.
>
> I think the one that Paul Crutzen and others suggested—putting sulphur dioxide into the stratosphere so it forms sulphuric acid droplets, a human-made volcano, in effect—[is an interesting idea]. You might say that's dangerous, because we don't know what's going to happen, and to some extent that's true even for Crutzen's

suggestion, but nature has performed that experiment. The Mount Pinatubo eruption in 1991 is interesting because it was large enough that for one year—that one year after the eruption—it was reflecting back to space about four watts of energy per square metre.

That's [cancelling out the equivalent of] doubled carbon dioxide, [the] equivalent of 560 parts per million of carbon dioxide. It's a big negative forcing.

One interesting point about it is that if you look at the melting in Greenland for the period when we have data, which began with satellite measurements in 1978–79, until the present, so thirty years, the year with the least melting was 1992, when those aerosols had maximum optical thickness. The sunlight has to go through them at a slant angle to hit this high-latitude ice-sheet, and it reflected enough sunlight away that it minimised the melting. So if the concern becomes especially these ice-sheets and their impact on sea-level, then you may have to seriously consider that. But frankly it makes more sense to reduce the forcing that's causing the problem.

—James Hansen, Director, NASA Goddard Space Studies Centre, in an interview with the author, June 28, 2008

There is good reason to be cautious when it comes to geo-engineering, because the moral hazard is real. Prolonged reliance on geo-engineering techniques to keep the global temperature stable while carbon dioxide concentrations continue to rise would probably have catastrophic effects on life in the oceans, which are already becoming distinctly more acidic due to a higher level of dissolved carbon dioxide in the water. (In the water, carbon dioxide turns into carbonic acid, which is what rots your teeth when you drink too many soft drinks.) Too much acidity, and the tiny marine animals that are at the bottom of the food chain have more and more difficulty in forming their shells.

As a very long-term solution to the problem of global warming, any kind of geo-engineering is a time bomb, because over time, as more and more carbon dioxide accumulates in the atmosphere, the difference between the actual average global temperature, artificially lowered by geo-engineering, and the temperature that the existing concentration of carbon dioxide would normally imply grows wider and wider. All of the techniques under discussion require continuous supervision and continual inputs of energy and materials from a highly organized society. If a major war, plague or other event causing social breakdown should lead to an interruption in the service, then there would be a sudden jump in the global temperature of three, four or even five degrees (depending on how long the dependence on geo-engineering had been), and an equally sudden collapse of almost all agricultural systems. Hardly any domesticated plants could withstand a sudden shift in temperature on that scale .

But nobody is actually proposing geo-engineering solutions as an alternative to cutting greenhouse-gas emissions. They are generally seen by their advocates as temporary devices for winning time and keeping the temperature from rising into the dangerous zone of uncontrollable feedbacks *while work continues to get emissions down to a safe level.* As for the moral hazard involved in discussing these techniques in front of the children, it's too late. The children already know. Indeed, the danger is that they may have too much faith in the ability of geo-engineering techniques to save them from their own folly.

> The thing that most of us want to see happen is to cut emissions quickly enough that the climate change is small enough that we just live with it, but the actual amount of climate change we get for a given amount of emissions is still deeply uncertain. It's been uncertain for forty years, since President Johnson got the first high-

profile report on the climate problem when I was a boy, and it's still uncertain. We're not going to know how big the climate threat is until after it's too late; until after we've put the carbon dioxide into the air.

Say it's 2050, and maybe we've actually done a great job of cutting emissions, and maybe emissions are well on their way down at that point, and concentrations [of carbon dioxide] have peaked at some level, but we still find that we're melting the Greenland ice sheet, which is something that might happen. Then we might find that we wanted to stop it. That would not be using the techniques of geo-engineering *instead of* cutting emissions; it would be *as well as* cutting emissions, to minimize the worst effects . . .

The biggest concern with geo-engineering—whether or not we should talk about it, whether or not we should do any research now—is that the mere knowledge that it could be done will reduce the incentive to cut emissions now. The short answer is simply that the cat is out of the bag. You can't make good policy by hiding your head in the sand. At this point, what we have is a sort of blogosphere around geo-engineering, with an enormous number of people talking about it, and very little in the way of good research. It might be that none of the geo-engineering techniques we have actually work . . .

Let's be clear: none of these schemes are an exact compensation for not emitting carbon dioxide. All of these schemes we have for engineering the planetary heating balance, if you like, will have some negative consequences. Perhaps very serious ones, perhaps not so serious. So I think a research programme aimed at understanding how we do it and what the conse-quences are is better than the current situation, where we agree that it's politically incorrect to talk about it, but everybody sort of knows it can be done and perhaps overestimates how much it can be done. I think that's more dangerous than real knowledge.

—David Keith, Canada research chair in
energy and the environment, University of Calgary,
in an interview with the author, May 2, 2008

The avalanche was waiting to happen anyway, but it was Paul Crutzen's article in 2006 that set it in motion, and now there are half a dozen different proposals on the table that could reasonably be called geo-engineering. Most of them fall into two main categories: fast-acting techno-fixes that directly reduce the amount of solar energy reaching the Earth's surface, and slower processes for extracting carbon dioxide from the atmosphere. On the assumption that you can handle this knowledge without immediately emitting immense amounts of carbon dioxide, I will attempt a brief survey of the field.

Crutzen's proposal was far from the first time that somebody had suggested mimicking the action of volcanoes, which also inject large amounts of sulphur into the stratosphere, but he brought two new things to the table: a rough costing of the enterprise, which he reckoned would be equivalent to 5 or 10 percent of the current U.S. defence budget, or US$25–50 billion annually, and some tentative reassurance on the question of whether doing this on a long-term basis would damage the ozone layer. While sulphur dioxide does not directly attack ozone, in the presence of chlorine compounds (derived from the chlorofluorocarbons [CFCs] that caused the ozone holes in the first place), it acts as a catalyst and speeds up the destruction of the ozone in the stratosphere. However, Crutzen estimated, by the time that anybody would feel driven to resort to this kind of emergency measure, decades from now, the amount of CFCs in the upper atmosphere will have fallen to levels that do not significantly endanger ozone concentrations, even if sulphur dioxide is present as well. The Montreal Protocol of 1987 has done its job, and CFC emissions are almost at an end now.

Since Crutzen's Nobel Prize was directly related to his work on ozone chemistry in the stratosphere, his estimates carried considerable weight in the scientific community. Moreover, his alternative, less articulated proposal for dispersing elemental carbon (soot) in the stratosphere as a coolant actually promised to raise stratospheric air temperatures, thereby inhibiting the formation of the ice-crystal clouds where massive amounts of ozone get destroyed in the late winter each year. Others have subsequently pointed out that the sulphur dioxide need not be sent up into the stratosphere by the cumbersome means of balloons (resembling giant weather balloons, presumably) or artillery. It could also be delivered to the stratosphere simply by dosing jet aircraft fuel (avgas) with a 0.5 percent solution of sulphur, which would lower the cost of the operation by an order of magnitude. This was the method favoured by Professor Tim Flannery of Macquarie University in Sydney, leading climate scientist and Australian of the Year 2007, who observed in May 2008 that climate change was moving so fast that it might be necessary to start doing this within five years.

The sulphur-in-the-stratosphere idea, which has the legitimacy of mimicking a natural process and requires no new technology, has a huge lead over all rival proposals for cutting incoming solar energy. The others all envisage some form of physical sunshade, whether by spraying a myriad of tiny, shiny balloons into the stratosphere or boosting a huge cloud of light-refracting mini-spacecraft (sixty centimeters in diameter, weighing about a gram each) into a stable orbit between the Earth and the Sun. The latter is intriguing next-generation-but-one technology — and if technological hubris were a disease, then Roger Angel, director of the Center for Astronomical Adaptive Optics at the University of Arizona, would probably be dead.

Angel's idea is to insert his fleet of sixteen trillion gossamer-light spacecraft into the L1 or Lagrange point about 1.6 million kilometres from Earth along the Earth-Sun axis. This is in some ways the equivalent of a geostationary orbit, except that objects put into solar orbit at the L1 point remain directly between the Earth and the Sun without any further expenditure of energy. Sixteen trillion spacecraft sounds like rather a lot, but each one is basically a transparent film less than a metre in diameter, pierced with small holes, that costs very little to produce. Each would weigh about as much as a butterfly, and they would be fired into space in stacks of a million by electromagnetic launchers installed on a mountaintop near the equator. A total of twenty such magnetic railguns, launching a stack of the flyers every five minutes for twenty years, would give Angel his sixteen trillion flyers, which, on arrival at the Lagrange point, would be dealt off the stack into a large, cylindrical cloud with a diameter about half that of Earth but ten times longer.

The cylinder would be oriented lengthwise along the Earth-Sun axis, so that much of the sunlight destined for Earth would pass through a sixty-thousand-kilometre-long cloud of transparent flyers. Each flyer would divert about 10 percent of the light hitting it away from the Earth. The rest of the light would pass through and continue to Earth, while the slight pressure exerted by the solar radiation, manipulated by tiny mirrors that act likes sails, would enable each flyer to maintain its position in the cloud (which would not be that dense, really: the average distance between the tiny flyers would be about a kilometre). Larger, unmanned control craft would send the orders that kept the cloud of flyers in its cylindrical shape.

The net effect would be to reduce sunlight by about 2 percent over the entire planet, enough to counterbalance the heating caused by a doubling of atmospheric carbon dioxide in the Earth's atmosphere. If circumstances on Earth changed and more solar energy was needed, however, the control craft could also let almost all the sunlight through simply by moving the flyers out of their cylindrical formation and spreading them out in a flat lens.

It is wonderfully ambitious technology, and Angel calculates that it could be done in a quarter-century for about a trillion dollars. Like all the people working in geo-engineering, he sees it as a last-ditch proposal: In a *Guardian* interview on May 29, 2008, he said that the "potential side-effects of geo-engineering and the cost of doing it in space would be inhibitors to doing this unless we felt desperate." The only drawbacks are that it costs about forty times as much as Crutzen's proposal, takes twenty-five times as long to complete, and depends on currently unavailable technology. On the other hand, it could be just what the twenty-fourth century needs (if there is a human twenty-fourth century), because if they decide to terraform Mars and Venus, this is probably the most efficient way of lowering the temperature on Venus and raising it on Mars.

Meanwhile, back here on Earth, there are various proposals on the table that would simply remove large quantities of carbon dioxide from the atmosphere. The cheapest and most effective, of course, would be massive reforestation projects in the tropics (deforestation currently accounts for an estimated 20 percent of annual carbon emissions by human beings), but since this is an intensely difficult issue politically, more technological projects for sequestering carbon are also begging for attention. The most successful, at least in terms of attracting public attention, has been the notion of fertilizing the ocean surface with scarce trace elements that normally limit the growth of marine micro-organisms.

In the case of both Climos and Planktos, two California-based start-ups, the scarce element on which they founded their business plans was iron. Very small amounts of this mineral are essential to the growth of the phytoplankton (microscopic plants) that are both the foundation of the oceanic food chain and also, if they die uneaten and sink to the bottom, the main vehicle by which carbon is sequestered in the ocean depths permanently or for very long periods of time. Runoff from the land tends to provide coastal seas with adequate amounts of iron, but in the open ocean basins, the only means of delivery is sporadic deposits of wind-blown dust. Many species of phytoplankton have evolved to take rapid advantage of these occasional bonanzas of iron-rich dust, and hence the brief phytoplankton blooms, lasting about sixty days, that occur in the oceans from time to time.

It was argued that changes in agricultural practice had seriously reduced the amount of wind-blown dust over the oceans since 1980, and that NASA and NOAA (National Oceanic and Atmospheric Administration) studies showed a corresponding decline of 25 percent in Pacific plankton populations since that time (6–9 percent globally). What these companies proposed to do was to replace some of that missing iron by sowing finely powdered iron dust across large tracts of open ocean — Planktos suggested that fifty tonnes would be adequate for ten thousand square kilometres — in order to artificially stimulate phytoplankton blooms. The phytoplankton would absorb enormous amounts of carbon dioxide as they grew, and those that survived to the end of their sixty-day life-cycle would then sink to the bottom, carrying megatonnes of carbon down with them. Even those that were eaten by zooplankton (the microscopic animals that multiply almost as fast during a bloom) would serve a useful purpose, as the zooplankton would in turn be eaten by crustaceans and fish, enhancing the biomass of the entire oceanic region.

The great bulk of global carbon fixation takes place in the oceans. Planktos talked grandly on its website about creating "forests in the oceans," and suggested that "ocean iron fertilization (OIF)," as the technique is now known, might eventually restore the annual three gigatonnes of carbon sequestration by phytoplankton that had been lost since 1980 because of the changed pattern of wind-borne dust. If that were true, it would be quite useful, as it is equal to approximately half the current emissions from all human industrial and transport activities in a year. Indeed, the Planktos website boasted that "Returning plankton populations to 1980 levels would neutralize about 50%

of industrial society's greenhouse gas emissions, and we feel that is about all you can or should ask a single ecosystem to contribute to our self-inflicted climate wars."

You will, of course, be wondering where the profit lies in all this (for these companies are financed by venture capital). The answer is that if they could demonstrate that their technique actually generated large phytoplankton blooms that would not have occurred naturally, and if they could show that this did no harm to other chemical or biological processes in the ocean, and if they could prove that the net result of the bloom is to remove substantial amounts of carbon dioxide from circulation for a long time (more than a hundred years), and if they could document and measure just how much carbon dioxide they had moved to the ocean bottom, *then* they could sell that amount of sequestered carbon dioxide as carbon credits. Rather a long string of ifs, and profitability also depends on a global cap-and-trade market that provides a decent price for carbon. (Dan Whaley, founder and CEO of Climos, estimated in 2007 that the cost to the company would be three to seven American dollars per tonne of carbon dioxide that was sequestered.)

As with most geo-engineering schemes, however, experiments with OIF have not been large enough or long enough to yield conclusive answers about the utility, cost and environmental implications of the technique. Much powdered iron has been thrown off the sterns of ships, and local phytoplankton blooms have duly occurred, but the studies that would provide real data, for example, about what proportion of the phytoplankton would sink to the bottom, removing carbon dioxide from circulation, and how much would simply be eaten, have not yet been done. Presumably because the money hasn't been there to do them, and it still isn't.

Indeed, in 2008, Planktos announced that it was pulling the plug on further ocean experiments, called its ship home, and laid off most of its employees. Climos, which has managed to attract some very distinguished scientific names to its advisory board, continues to express optimism about the long-term prospects for the technique, but it is now effectively hamstrung for some time to come. In May 2008, the ninth conference of the 191 countries that have signed the UN Convention on Biological Diversity responded to the demands of a coalition of environmental groups by calling for a moratorium on the practice of adding nutrients to the ocean. Although the moratorium has no legal force, it is a huge deterrent to any further experimentation for the time being.

A different approach to ocean fertilization was taken by the Australian firm Ocean Nourishment Corporation, which proposed on its website to "increase the amount of carbon involved in the naturally occurring organic carbon cycle, through the enhancement of phytoplankton stocks in nitrogen deficient regions of the world's oceans." In these nitrogen-poor areas, mainly near the edges of the continental shelves, the company proposed to release a modest but steady flow of highly diluted urea, a nitrogen-based fertilizer, from underwater pipes in order to encourage the growth of phytoplankton. Not enough would be released to cause algal blooms—only about one-fifth to one-tenth as much—but it would allegedly be enough to support denser populations of phytoplankton, and therefore of other ocean species all the way up the food chain. Overall biomass would increase, and so would the absorption of carbon dioxide and the amount of carbon ultimately sequestered on the bottom. Having documented this process, Ocean Nourishment would then claim carbon credits for the extra carbon dioxide removed from the atmosphere.

Ocean Nourishment's first experiment was to be carried out in the sea between the Philippines and Borneo in 2008, but like Climos it felt compelled to suspend operations by the call for a moratorium. The issue has now been referred to the London Convention, the body that regulates the dumping of wastes at sea and, at the time of going to press, the outcome is uncertain.

In the absence of any legislation that covers this sort of activity, it was imaginative of the environmentalists to redefine the experiments as the dumping of waste, an activity that is covered by legislation. Whether seeking to suppress these experiments is the right thing to do is quite a different question, and one whose answer is not clear. Intuitively, one suspects that there are more cost-effective and less invasive ways to sequester carbon, and everybody knows that the companies seeking to do these experiments are in it for the money. But intuition is not always a good guide in scientific matters, and it could be useful to know more about what is or is not possible.

A good example of the shortcomings of intuition is the seemingly preposterous idea of machines that would draw carbon dioxide out of the air directly—"artificial trees," as somebody called them. Given how low the concentration of carbon dioxide in the air is, it seems obvious that the energy requirements for separating it out would be so high that the machines would be hopelessly uneconomic. But, as David Keith explained to me in our May 2008 interview, that is not necessarily so:

> I've ended up, along with several different groups around the world, working relatively seriously on the idea that we could build industrial equipment to pull carbon dioxide directly out of the air, [to] make concentrated streams of carbon dioxide. At first you might think that this is an absurdly hard thing to do, because the amount of carbon dioxide in the air is a half of a part per thousand, whereas the amount of carbon dioxide coming out of a power-plant exhaust is more like 10 or 15 percent, and we're having a lot of trouble and spending a lot of money to do that. But surprisingly, it's not necessarily that much harder when you look at the basic thermodynamics.
>
> So there's been a fair amount of interest in this topic, and a guy called Klaus Lackner in the U.S. has been one of the leaders, and we're actually in the middle of patenting some technology that may produce competitive ways of doing this. We had this idea that the business

model would be that you suck carbon dioxide out of the air, put it underground and people would pay you in carbon credits. But when you actually go talk to real businessmen, they say: We don't want to put this carbon dioxide underground. We want to make products with it.

Timing is everything. At this point in the interview, we were interrupted by a phone call from the university parking lot, where a semi-trailer had just arrived bearing the prototype of David Keith's machine. Where did he want it unloaded? So we went downstairs, and while he dealt with the logistics of the situation, I had a look at the machine, which took up much of the interior of the trailer. All I can tell you is that it was a rather convoluted hunk of metal, painted green, and looking vaguely as if it had been designed by Rube Goldberg. David Keith wouldn't tell me how it worked, because the patent is still pending, but he was quite happy to talk about how the economics of extracting carbon dioxide from the air might work:

Let's say you are Richard Branson, and you run an airline, and you're worried that your airline will be shut down because of constraints on carbon dioxide. You know that you can't make electric airplanes or hydrogen-powered airplanes very well, and you think biofuels aren't going to work very well. One of the things you might want to do is sell truly carbon-neutral airplane seats, and air capture is a way to do that. Either by still burning petroleum in your airplane, but then sucking out an exact equal amount of carbon dioxide from the air and putting it underground in a certified way, so everybody agrees that you haven't added to the carbon dioxide in the air, *or* [by sucking] carbon dioxide out of the air and turn[ing] it into a fuel by adding energy—there's no free lunch here—and that might be far more cost-effective than doing hydrogen.

Let's say you have a source of hydrogen, which doesn't create carbon dioxide emissions, and you propose to use that to solve the global transportation problem. Forget where the hydrogen came from: could be solar power, nuclear power, not my problem right now. One option I have is the option we're all familiar with: that I put it in little hydrogen tanks in my car, and I run fuel cells and all that. There turns out to be immense problems with this, because hydrogen is a very low-density fuel with lots of problems . . .

Another, quite different option is you take the hydrogen, you capture carbon dioxide out of the air, you mix the hydrogen and the carbon dioxide, which quite easily makes a fuel—you can make octane if you want to—and then you sell a carbon-neutral hydrocarbon. So you sell something that is compatible with the existing vehicle fleet of the world, but the carbon in the stuff you're selling—which is octane, just like gasoline—is stuff you got from the air. It's like returning the empty bottles after they've drunk it. So you're selling them this fuel, and they're burning it, putting carbon in the air, but then you're recapturing the same amount of carbon and selling it to them again. That's a business model that could conceivably take a whack at the global transportation market, which is the hardest part of the climate problem to crack.

The rival team, led by Klaus Lackner, a physicist at Columbia University, has some numbers to offer. They estimate that their machine (about the same size as Keith's) would initially cost about US$200,000 to build, though that would come down with high-volume production. It would be able to "scrub" about a tonne of carbon dioxide a day from the atmosphere (about the same amount that would be emitted to carry one passenger on a full plane from London to New York), so five or six hundred of the machines could compensate for the carbon dioxide emitted by one airliner, assuming that it crosses the Atlantic twice in a working day. Put that way, it sounds ludicrously uneconomic, but it may not be: if the only alternative was to stop flying, Richard Branson might be able to build the cost of buying and operating them into the ticket price without driving all the passengers away.

The question comes down to money, in the end. How much power, at what cost, does it take to scrub that tonne of carbon dioxide out of the air? If it is a cost per tonne that is lower than the going price of a carbon credit for one tonne (and you have somewhere to sequester the carbon dioxide), or if some manufacturer of fuel is willing to pay more for a tonne of carbon dioxide than it costs to extract, then this is a viable technology. Is it also an occasion of sin? Of course it is, but against the accusation of moral hazard, Lackner argues: "We are in a hurry to deal with climate change and will be very hard pressed to stop the train before we get to 450 ppm [parts per million]. This can help stop the train. I'd rather have a technology that allows us to use fossil fuels without destroying the planet, because people are going to use them anyway." (*Guardian* interview on May 29, 2008)

Could it be done on a scale that would make a really big dent in the rapid and accelerating accumulation of carbon dioxide in the atmosphere? That seems hard to believe, since the human race is now adding six gigatonnes a year to the load—but if the scrubbers performed as promised, it would only take twenty million of them to counter all that carbon dioxide. That's about three times the number of large trucks (more than two axles or more than four tires) in the United States: a big number, but not an unimaginable one. A harder question, as in the case of CCS technology, is what to do with all that carbon dioxide, as twenty years' worth of human carbon dioxide emissions, captured and liquefied, would fill Lake Michigan.

It is worth doing these back-of-an-envelope calculations about sulphates in the stratosphere, oceanic iron fertilisation, and "air capture" of carbon dioxide, even if the geo-engineering technologies under discussion are largely untested, because of the very serious possibility that all the conservation and mitigation measures that people around the planet take in the next twenty or thirty years will not be enough to stop global warming short of the point where the feedbacks are activated in a major way. "Some people say we can stop at 450 parts per million, but that's ludicrous," said Wallace Broecker, Lackner's chief scientific collaborator, in a recent interview published in the *Guardian*, May 24, 2008. "It will be hard, very hard, for us to stop at even 600 ppm. And if we carry on doing what we are doing now—very little—we are going to get up to 800 or 900 ppm."

Relatively slow-acting techniques for drawing carbon dioxide out of the atmosphere, if they prove to be environmentally sound and economically viable, could be deployed even before the world faces an imminent and grave climate crisis, in order to slow the arrival of that crisis or perhaps even avert it. Drastic measures for holding the temperature down by blocking solar radiation, on the other hand, should obviously not be deployed unless a grave crisis is actually upon us.

In the case of putting sulphur dioxide into the stratosphere, delaying until the last minute is cost-free because the cooling effect is relatively fast-acting. You would not wish to do that in anything but an extreme emergency in any case, as it could entail a substantial environmental cost. But there is another proposal for cutting solar radiation that involves intervention not in the stratosphere but about two hundred metres above sea level. John Latham of the National Center for Atmospheric Research in Boulder, Colorado, explained it to me in a May 7, 2008 interview:

> The idea is to make clouds over the ocean reflect more sunlight than they do. I first started working on this idea in about 1990, but I think the stimulus for that came two decades before on a Welsh mountainside, looking out over the Irish Sea. My eight-year-old son Mike was with me, and we were looking at a beautiful sunset, and it was

very shiny in patches. We were above the level of the clouds, and he asked me why they were so shiny. So I explained what was going on, and he laughed and he said "Clouds are soggy mirrors."

There's a particular type of cloud, more extensive than any other clouds on Earth, called marine stratocumulus clouds. They're very low level, just a few hundred metres from the ocean's surface, they're very thin, just a few hundred metres thick, and they cover about a quarter of the oceanic surface. They reflect sunlight back to space—about 50 percent of the sunlight that lands on the top of those clouds is bounced back into space—so those clouds actually produce a very substantial amount of cooling naturally. But if we could rack up that percentage from about 50 to 55 percent, then we could produce a cooling that would be sufficient to compensate for the warming that results from a doubling of the carbon dioxide concentration in the atmosphere. So the idea is to make these clouds more reflective to incoming sunlight.

It isn't a perfect solution; nothing is. If you have lost the ice cover on the Arctic Ocean and the Earth's heat balance is sliding even further out of balance because the dark water is absorbing a lot more heat, perhaps you could compensate for that by seeding enough stratocumulus clouds with water droplets and reflecting an equivalent amount of sunlight. Unfortunately, you would not be cooling the part of the planet that most needs it, the Arctic, because the stratocumulus clouds are mostly found fairly close to the equator. You could end up with the worst of both worlds: huge disruptions to weather patterns in the equatorial regions, where you are making it cooler, while heating in the Arctic proceeds apace. You'd never choose to introduce such disruptions to the system voluntarily—but if the Arctic ice is already gone, maybe what you have to do to stop the Greenland ice cap from following suit is to concentrate on cooling the subtropical parts of the North Atlantic Current ("Gulf Stream"), which will later end up around Greenland. But don't cool it too much, or you might stop the whole current, which would give the Western Europeans a lot more cooling than they bargained for. Tricky business, this geo-engineering.

Latham's idea was partly chosen for its relatively low and controllable impact on the Earth climate system. He proposes fleets of unmanned, satellite-controlled vessels that spray tiny droplets of seawater into the air below the stratocumulus clouds. It doesn't take a lot of power: you just have to get the droplets a couple of metres up into the air, and updrafts will carry about half of them on up into the clouds a few hundred metres above. Once there, they increase the reflectivity of the cloud, which is determined mainly by the size of the droplets it contains. Smaller droplets are more reflective (because the same amount of water distributed among more and smaller droplets has a bigger surface area). Since the droplets that the ships spray would be considerably smaller than the average size of those that form naturally within the cloud, they would raise the reflectivity of the cloud by about 5 percent—and, as a bonus, clouds containing these smaller droplets tend to last longer before they disperse.

Stephen Salter, a professor of engineering design at Edinburgh University, has designed the wind-powered ships that would do the spraying. They would be relatively small— about forty metres in length and displacing only about two hundred tonnes, the size of a large ocean-going yacht—and could be mass-produced at a modest cost. They would have not conventional sails but Flettner rotors that generate thrust to move the ships forward by rotating in the wind (and also allow the ships to tack closer to the wind than fabric sails). The electrical power that rotates the cylinders and sprays the seawater into the air would be generated by very large propellers in the water that act as turbines, so the ships would need no fuel and could remain independent of land support for long periods of time.

Everybody working in the field of climate recovery would prefer solutions to the global warming problem based on the use of non-carbon sources. But fourteen years after the Kyoto agreement was ready for signature, the rate of increase of atmospheric carbon dioxide has itself increased. It is therefore urgent to design and test all possible measures to stabilize temperature, for use if the present proposed methods are unsuccessful.

It is technically possible to increase the reflectivity of marine stratocumulus clouds by spraying quite small quantities of sea water into the marine boundary layer . . . On reasonable assumptions of the present concentrations of condensation nuclei in mid-ocean air and drop life, the first global reduction of one watt per square metre will require spraying a total of only five cubic metres a second. Because of diminishing returns, the increase from 2.7 to 3.7 watts per square metre, needed to stabilize temperatures despite a future doubling of carbon dioxide levels, will require an extra thirty cubic metres of sea water a second.

If necessary it would be possible to spray amounts sufficient to compensate for five watts per square metre over the entire earth surface. Operations in water en route to the Arctic will have secondary benefits in the restoration of ice cover and the reduction of methane release from Siberian permafrost. Operations may also be aimed at endangered coral.

— Stephen Salter and John Latham, "The Reversal of Global Warming by the Increase of the Albedo of Marine Stratocumulus Cloud," *The Engineer Online*

The fleets of spray vessels would be directed to stratocumulus-rich areas of the ocean by satellite commands, and migrate between the Northern and Southern hemispheres to follow the summer (the greater the solar radiation hitting the top of the clouds, the more efficient the process). They would avoid shipping lanes, anywhere with large numbers of icebergs, and sea areas that have land downwind that might suffer from altered rainfall as a result of their activities, but at least 80 percent of the world's oceans would be available for their operations.

Latham and Salter estimate that deploying fifty spray vessels with an expected service lifetime of twenty years and costing a few millions dollars each would be enough to compensate for one year's worth of global warming at the current rate of increase of carbon dioxide in the atmosphere. Deployment of a further fifty spray vessels each year could cancel out the heating due to that year's carbon dioxide emissions, so the annual investment would be relatively modest. If there should be unforeseen negative consequences of the technique, the spraying could be shut off at once, and conditions would revert to normal in a matter of days. And if the global climate should spin out of control, with rapid methane release or some other feedback kicking in sooner than expected, the number of vessels could be expanded rapidly to prevent a sharp rise in temperature.

It sounds almost too good to be true, and maybe it is. But it is sheer fecklessness to fail to investigate such possibilities aggressively, because we are currently conducting an unplanned experiment in global climate alteration through massive carbon dioxide release without any kind of safety net. It would be comforting to have at least one reserve position to fall back on, in case all those promises of future emissions cuts don't come true. You know, just like all the past promises of emissions cuts didn't come true. Three or four different tested and proven options for how to stop the temperature from soaring if the Kyoto process or its son or niece or second cousin doesn't deliver the goods in time would be even nicer.

Moral hazard be damned. This is serious.

SCENARIO SEVEN:
WIPEOUT

IF I WERE GOING TO WRITE THIS FINAL SCENARIO the way I did the other ones, I would put it in about 2175, I would set the average global temperature at nine degrees Celsius higher than the present (much higher in the polar regions, of course), and the scenario would start something like this:

It was a much simpler global society now: three hundred million people speaking only two major languages, English and Russian, clustered around the shores of the Arctic Ocean (although those shorelines, after seventy metres of sea-level rise, would have been unrecognizable to their great-grandparents). There were scattered habitable parts of the world further south, like the British archipelago, Newfoundland and the mountainous interior of British Columbia, and even a few coastal habitats in the tropics that still got enough rainfall to support a human population, but the interiors of all the continents were burning deserts. In the Southern Hemisphere, most of New Zealand was still densely populated, as was Patagonia, and there were a few attempts to create settlements on the shores of Antarctica, but that was it. Some people dreamed of massive geo-engineering projects that would bring the temperature back down and let the human race recolonize the rest of the planet, but the new society did not have the resources of the old days, and besides it had a more urgent problem. The oceans were going bad. Going bad in the sense that they smelled like rotten eggs.

But I'm not going to write that sort of scenario here. It would be too melodramatic, too apocalyptic. Less dramatic outcomes to the current climate crisis seem far more probable. We should be aware of where this might all end up if we really make a mess of the task before us, however, and most people aren't. Even many scientists aren't, because the evidence is quite new. So let us consider the new hypothesis about what caused all but one of the "Big Five" mass extinctions in the Earth's history.

The people who came up with this hypothesis are not climate scientists. They are paleontologists who study the deep past, and in particular those who focus on the emergence and extinction of animal and plant species over the past half-billion years, for which there is reasonably abundant fossil evidence. The only book-length popular account of the new mass-extinction hypothesis is *Under a Green Sky*, written by Peter Ward, one of the paleontologists who was prominent in the research and controversies that led to its formulation. I have relied heavily on his book, published in 2007, for the following discussion.

As early as the mid-nineteenth century it was understood that the history of life on this planet was punctuated by great extinction events during which a very large proportion of all living species died out in a relatively short period of geological time. Beginning in the 1840s, English naturalist John Phillips, counting the diversity of marine species present in the sediments from various eras, proposed that the history of life could be organized into three great eras, divided by huge mass extinctions. There was the Paleozoic era, the time of "old life," which ended with a great extinction about 250 million years ago; then the Mesozoic ("middle life") era, which concluded with another mass extinction around 65 million years ago; and, finally, the Cenozoic ("new life") era, which continues into the present.

Those major divisions of time remain in use even today, but subsequent generations of paleontologists have identified more extinction events, including a total of five very big ones, since organisms with skeletons appeared on the planet during the "Cambrian explosion" of 530 million years ago. (There may have been other mass extinctions before the Cambrian era, but the life of that earlier time left little fossil evidence of its presence.)

The Big Five mass extinctions, a category first suggested in 1982 and widely recognized by paleontologists today, were the end-Ordovician (444 million years ago); the end-Devonian (360 million years ago); the end-Permian (251 million years ago); the end-Triassic (200 million years ago); and the end-Cretaceous (65 million years ago). The very worst was the end-Permian or Permian-Triassic (P-T) event, sometimes called "the great dying," in which about 96 percent of all marine species and an estimated 70 percent of land species, including plants, insects and vertebrates, vanished from the fossil record. The most recent, the end-Cretaceous or Cretaceous-Tertiary (K-T) event killed only about 50 percent of all species, but it had great significance for us since it ended the age of the dinosaurs and bequeathed the land surface of the planet to mammals and birds. Since human beings live on the same planet where all these mass extinctions occurred, only a little bit later, the reason why they occurred is a matter of great interest to us. Until the 1980s, the dominant hypothesis held that they were all caused by gradual climate change.

"Hypothesis" may be too generous a term, because this was more an evasion of the question of why all those species died out than an explanation: it proposed no kill mechanism that could account for it. Why would gradual climate change—and everybody agreed that it was happening on the scale of thousands if not millions of years—cause mass extinctions? It's a big planet and there's plenty of time: each species needs only to migrate to wherever its preferred climatic conditions have moved.

The sheer illogic of the traditional hypothesis becomes clear if you just consider the record of the Current Ice Age, which began only 26 million years ago. Since then, there have been around twenty major glaciations when ice sheets covered much of North America and Eurasia and the global climate was dry, windy and about five degrees Celsius colder than at present, interspersed with briefer interglacials like the one we are in today. Those are very big changes in climate, happening repeatedly over relatively short periods of time, and yet there has been no mass extinction in this period (other, perhaps, than the one that human beings have begun to cause in the past couple of centuries). The animals and plants just moved back and forth in their preferred climatic bands as those shifted towards the poles or back towards the equator.

Nevertheless, the hypothesis of "gradual climate change" had been around for so long that it had become holy writ, and there was great resistance when the father-and-son team of Luis Alvarez (a physicist) and Walter Alvarez (a geologist) discovered that a clay layer that dated back right to the K-T boundary sixty-five million years ago was very rich in iridium, an element extremely rare on Earth but common in asteroids. In a famous paper in 1980, they suggested that the Earth had been struck by a giant asteroid sixty-five million years ago, and that that was what wiped out the dinosaurs. This set off a decade-long war between the older generation of paleontologists, most of whom rejected such a radical suggestion out of hand, and some young guns who were open to the possibility.

By 1982, the same tell-tale iridium had been found at the K-T boundary layer in forty different places around the world, together with "shocked quartz" grains showing multiple thin lines called "shock lamellae" that are found nowhere on Earth except in meteor impact craters and underground nuclear test sites. These were found together with large numbers of tiny, glassy, bead-like spherules that the pro-Alvarez scientists interpreted as material that had been blasted out of the Earth's atmosphere and all around the world by the impact, then melted with the heat of re-entry into their characteristic shape. In the same boundary layer they found fine particles of soot, which they saw as evidence of worldwide forest and brush fires.

Almost everybody agreed by 1982 that the planet had indeed been struck by a giant asteroid at least ten kilometres in diameter sixty-five million years ago. Eventually, in the early 1990s, they found the actual impact site, the 180-kilometre wide Chicxulub crater (long since filled by sediments) on the coast of Mexico's Yucatan peninsula, but the fact that there had been a massive impact with worldwide effects just at the time when the dinosaurs and so many other species became extinct did not prove that the old gradual climate-change hypothesis was wrong; the asteroid impact could be just a (rather large) coincidence. It took all the rest of the 1980s for young paleontologists to prove that the K-T extinction was not gradual at all, but a sudden event that devastated a thriving ecosphere.

They provided the proof by finding fossil beds containing the full range of species typical of the late Cretaceous right down to the K-T boundary. Then most of these species vanished suddenly and together: there had been no gradual die-off. By the early 1990s, the argument about the K-T extinction was settled — and the weight of paleontological opinion moved rapidly to the opposite pole: that all the great extinctions had extraterrestrial causes. There was no evidence to support this, but at least giant asteroid strikes gave you a kill mechanism you could believe in. So off went the researchers, searching for similar iridium-rich layers in the rocks at the level of the other Big Five mass extinctions.

They spent a dozen years looking for the evidence — and found nothing. The end-Ordovician, the end-Devonian, the massive P-T event, the end-Triassic: there was no evidence whatever of a large asteroid impact at the time the dying happened.

Moreover, unlike the K-T event, where the extinction was a single hammer blow that did almost all of its work in months or a few years, all these other mass extinctions seemed to have happened over some millions of years in a series of "pulses." There would be one very big loss of species that more or less defined the event, so they date the end-Triassic extinction to 199–200 million years ago, for example. But in fact there were also several lesser extinctions extending back for ten million years before that date and forward for several million years after it, as though something lethal kept coming back for another bite.

Eventually, however, two patterns did emerge that seemed to link these non-impact mass-extinction events. One was that they almost all seemed to be associated with elevated levels of carbon dioxide in the atmosphere. But that didn't solve the problem. So it was warmer; so what? How could that cause a mass extinction? But the other piece of evidence was that the deep oceans were anoxic (oxygen was very scarce or completely absent) at the time. Indeed, at the critical time when the extinctions occurred, they were "Canfield oceans."

It was known that the oceans in the deep past were often radically different from the oceans of today, which are rich in oxygen all the way down to the bottom. The ancient oceans were frequently stratified, with an oxygenated upper layer supporting familiar varieties of marine life, albeit in unfamiliar forms, and a lower layer that was very low in oxygen, supporting very few organisms. (One of the markers of that kind of ocean in the geological record is very smooth, undisturbed ocean bottoms, because all of the tiny burrowing animals that churn up the sediment in unstratified oceans are absent.) But geologist Donald Canfield, director of the Nordic Center for Earth Evolution and a professor at the University of Southern Denmark, was the first to point out that when the deep oceans became completely free of oxygen, the sulphur bacteria came out of the sediments where they hid from the oxygen and took over.

Producing hydrogen sulphide as a waste product of their metabolism, the sulphur bacteria made the deep ocean a deadly place for oxygen-based life. This deep, anoxic layer was separated from the oxygenated surface layer by a "chemocline" that was rarely more than a couple of hundred metres below the surface. The only large body of water like that in the world today is the Black Sea, where the chemocline is about 150–200 metres down, but in the deep past, Canfield argued, even the open oceans were in this state for long periods of time. He made his case, and they are now known as Canfield oceans.

Hydrogen sulphide is deadly not only to marine life. If it should make its way into the atmosphere in high enough concentrations, it is equally deadly to land animals and plants. By the beginning of the twenty-first century, most paleontologists had abandoned the hypothesis that all mass extinctions were caused by asteroid strikes, and they were looking for another kill mechanism that could explain them. The vague old story about gradual climate change no longer convinced anybody, so if it wasn't asteroids, what was it?

In May 2005, Lee Kump, of Pennsylvania State University, and two colleagues, Alexander Pavlov and Michael Arthur, published a landmark paper suggesting that Canfield oceans were the culprit. The paper, published in *Geology*, argued that once the concentration of hydrogen sulphide in the lower, anoxic layer got high enough, it could break through the chemocline, drive all the oxygen out of the surface layer, and escape into the air:

> Simple calculations show that if deep-water H_2S [hydrogen sulphide] concentrations increased beyond a critical threshold during oceanic anoxic intervals of Earth history, the chemocline separating sulfidic deep waters from oxygenated surface waters could have risen abruptly to the ocean surface (a chemocline upward excursion). Atmospheric photochemical modeling indicates that resulting fluxes of H_2S to the atmosphere (>2000 times

the small modern flux from volcanoes) would likely have led to toxic levels of H_2S in the atmosphere. Moreover, the ozone shield would have been destroyed, and methane levels would have risen to >100 ppm. We thus propose chemocline upward excursion as a kill mechanism during the end-Permian, Late Devonian, and Cenomanian–Turonian extinctions . . .

The Canfield oceans occurred at times when there was a heightened level of carbon dioxide in the atmosphere. That, in turn, could be explained by the enormous amounts of the gas that were released during large and long-lasting episodes of volcanic eruptions and basalt flooding like those that laid down the Siberian Traps at the time of the end-Permian extinction or the Deccan Traps in southern India at the time of the end-Cretaceous extinction. ("Traps," from the Swedish word for steps or stairs, are layered lava flows up to two thousand metres deep that cover huge areas. The Deccan Traps were originally about a million and a half square kilometres, and even sixty-six million years of erosion later they still cover half a million square kilometres) So we know where the carbon dioxide came from, and we know that it caused major global warming (though it took a long time to do so because the quantities released per year by the volcanoes were greatly inferior to those being released by industrial civilization today). But what ties global warming to the emergence of a Canfield ocean?

The oxygen in the lower depths of the oceans will be used up over time if it is not continuously replenished by descending currents that bring fresh, oxygen-rich water down from the surface, because the constant rain of dead organic matter from above uses up the available oxygen in the lower oceans as it decays. Ever since continental drift put the continents into their current positions, these descending currents have been located in the northern North Atlantic (at the northern end of the "Gulf Stream") and in the Southern Ocean between the South

Atlantic and Antarctica. Together, these currents are known as the Meridional Overturning Circulation (MOC) and, as they sink, they deliver huge volumes of cold, oxygenated water to the abyssal depths of the Atlantic Ocean, from which deep-water currents carry the same water into the Indian and Pacific Oceans. And as long as the points where the currents descend into the depths are in the high latitudes, all is well and the oceans stay oxygenated all the way down.

If major global warming occurs, however, the surface currents may start to descend ("overturn") well short of their present high-latitude destinations. The North Atlantic current seems particularly prone to this behaviour; indeed, it is known to have stopped entirely for several centuries-long periods during the rapid warm-up from the last major glaciation between ten and fifteen thousand years ago. A few years ago, researchers were concerned that a repetition of this behaviour might again lead to rapid and drastic cooling over Western Europe, which derives much of its heat from the warm surface waters of the North Atlantic current, but more recent data suggest that there is as yet no significant drop in the speed and volume of the current. It will take more warming than we have had yet to stop the current from reaching the high latitudes—but in the deep past there has been more warming, and it did begin to overturn a long way south of its present terminal points up in the seas between Greenland, Iceland and Norway. And when that happened, the ocean depths started to become anoxic.

The fact that the currents of the Meridional Overturning Circulation normally carry huge amounts of oxygen down to the lower oceans when they overturn is what allows oxygen-breathing marine life to flourish all the way down to the bottom, but that is a happy by-product of the MOC's real function, which is to equalize the different densities, temperatures and salinity levels that arise in different parts of the oceans. If the currents should start overturning well short of their present terminal points, then they will carry not cold, highly oxygenated water down to the lower depths, but warmer water containing much less oxygen. If the amount of oxygen coming down falls short of the amount that is being used up by decaying organic matter, then the deep oceans begin to go anoxic—and if the process continues to the point where there is no oxygen left, they turn into Canfield oceans.

Nobody knows how long this process takes, but it would probably be quite a long time. The deep-ocean currents move slowly and the volumes of water involved are immense. Moreover, no one knows how long it takes for an anoxic lower ocean to fill up with hydrogen sulphide once the relevant bacteria emerge from the silt and go into action, nor what concentration of hydrogen sulphide would cause the chemocline to rise to the surface and start emitting the gas into the atmosphere. Intuitively, it doesn't feel like an imminent danger with even the maximum amount of warming that is anticipated for this century—but this is an area where intuition is not a good guide. You can't have a "feel" for this stuff, because it's all unexplored territory.

What would it be like if a Canfield ocean had a "chemocline upward excursion," as the scientists put it, and began to emit large volumes of hydrogen sulphide? Almost all marine life would already be dead, of course, right down to the plankton. According to Kump's calculations, there would be more than enough hydrogen sulphide bubbling out of the oceans to kill off most land life as well, either directly by poisoning it, or by destroying the ozone layer and exposing animals and plants to lethal ultraviolet radiation from the sun. It has happened many times before, and it cannot be excluded that it might happen again. Peter Ward tries to imagine what the world would look like during such an episode in his book, *Under a Green Sky*:

> No wind in the 120-degree [Fahrenheit] morning heat, and no trees for shade. There is some vegetation but it is low, parched, stunted. Of other life, there seems little. A

scorpion, a spider, winged flies, and among the roots of the desert vegetation, we see the burrows of some sort of small animals—the first mammals, perhaps . . . The land is a desert in its heat and aridity, but a duneless desert, for there is no wind to build the iconic structures of our Saharas and Kalaharis. The land is hot barrenness.

Yet as sepulchral as the land is, it is the sea itself that is most frightening. Waves lap slowly on the quiet shore, slow-motion waves with the consistency of gelatin. Most of the shoreline is encrusted with rotting organic matter, silk-like swaths of bacterial slick now putrefying under the blazing sun . . . We look out on the surface of the great sea itself, and as far as the eye can see there is a mirrored flatness, an ocean without whitecaps. Yet that is not the biggest surprise. From shore to the horizon there is but an unending purple color—a vast, flat, oily purple, not looking at all like water, not looking like anything of our world. No fish break its surface, no birds . . . dip down looking for food. The purple color comes from vast concentrations of floating bacteria, for the oceans of Earth have all become covered with a hundred-foot-thick veneer of purple and green bacterial soup.

At last there is motion on the sea, yet it is not life, but anti-life. Not far from the fetid shore, a large bubble of gas belches from the viscous, oil slick-like surface, and then several more of varying sizes bubble up and noisily pop. The gas emanating from the bubbles is hydrogen sulfide, produced by green sulfur bacteria growing amid their purple cousins. There is one final surprise . . . High, vastly high overhead are thin clouds, clouds existing at an altitude far in excess of the highest clouds found on our Earth. They exist in a place that changes the very color of the sky itself. We are under a pale green sky, and it has the smell of death and poison.

The particular "greenhouse extinction" Ward is imagining here took place two hundred million years ago: the end-Triassic extinction. But his argument, and that of many other paleontologists now, is that this has happened not once but many times during the history of life on this planet. It's a simple causal chain. First, the world's climate warms rapidly because of a sudden increase of carbon dioxide and methane from massive volcanic eruptions accompanied by basalt flooding. In this warmer world, the ocean circulation systems change and the Meridional Overturning Currents of the Atlantic start sinking into the depths at lower latitudes, dumping warm, low-oxygen water into the ocean abyssal plains. Warming continues, and the difference between temperatures at the equator and the poles shrinks until there is almost no wind and hence no currents. The oceans by now have separated into two layers that do not mix: well-oxygenated water near the surface, and a bottom layer that is becoming more and more anoxic. Moreover, the bottom layer is expanding at the expense of the upper layer.

Finally the upper boundary of the anoxic bottom water rises to a depth where light can penetrate, and a combination of low oxygen and light allows green and purple sulphur bacteria to expand in numbers until, one fine day, the chemocline rises to the surface and they own practically the whole ocean. Most of the remaining marine life dies at once, but presumably some isolated areas are not affected, because some marine species make it through the extinction.

The bacteria, now thriving in the sunshine that penetrates the shallow water, produce vast amounts of toxic hydrogen sulphide: as much as two thousand times what volcanoes put into the atmosphere today. The gas rises into the high atmosphere, where it breaks down the ozone layer, and land animals and plants begin to die from the unfiltered ultraviolet radiation that is reaching the surface. Invisible clouds of hydrogen sulphide drift inland, poisoning plants and animals alike, the hydrogen sulphide even more effective as a killer because of the high heat. All this happens not with a bang, but a whimper. And it has happened not once, but perhaps twenty or thirty times over the past half-billion years. This time around, however, we are the volcanoes.

Two consoling thoughts. First, this is all a hypothesis, still open to refutation by contrary evidence. That would be nice. Unfortunately, it's a pretty solid hypothesis that has rapidly gained broad acceptance among paleontologists.

Second, even if the hypothesis is true, mass extinctions caused by Canfield oceans have not occurred for some time now. There appear to have been several dozen of them between 490 million and 93 million years ago, some global in extent and killing more than half of all the species then alive in the sea and on the land, others more regional and limited. (Not all the world's oceans necessarily enter the Canfield state at the same time.) But there has only been one occasion since then, about fifty-five million years ago, when there was a minor greenhouse extinction, affecting mainly marine species, that appears to have been caused by a Canfield ocean. Something may have changed that makes this kind of disaster less likely now, and one obvious candidate is the fact that the carbon dioxide concentration in the atmosphere is now far lower than it was in Permian or even in Cretaceous times: the previous catastrophes generally started from much higher temperatures.

On the other hand, that last greenhouse extinction occurred when there were only about eight hundred parts per million of carbon dioxide in the atmosphere, a level we might well achieve in this century on a "business as usual" basis. (In the whole of the nineteenth century, human activities added about 15 parts per million of carbon dioxide to the atmosphere; during the twentieth century, we added 75 parts per million, bringing the concentration up to 370 parts per million, and we have already added another 18 parts per million since then. At the moment, we are adding almost 3 parts per million per year and the rate is still rising; we could easily hit 700 or 800 parts per million by the end of this century.) Would that level of warming initiate the changes in the Meridional Overturning Circulation that would start the world down the road to a Canfield ocean?

Not necessarily. There were huge and abrupt fluctuations in temperature many times during the hundred thousand years that preceded the current 10,000-year era of almost eerie temperature stability, but the impact of those rapid swings on the MOC has been different during this ice age than it was in the deep past. The North Atlantic current has completely shut down on numerous occasions, bringing an extra wave of cold to northwestern Europe, but it did *not* reorganize itself so that the overturning process moved further south and started delivering a torrent of relatively warm, low-oxygen surface water to the bottom. Rather it just stopped entirely for a while (which did deprive the lower oceans of oxygen for that period, of course), and then resumed its previous flow when circumstances improved. The system has a momentum that is not easily diverted and, as a result, the oceans have remained well-mixed from top to bottom.

The pattern of the deeper past, linked by the new hypothesis to the greenhouse extinctions, was that when global warming struck, the MOC continued to run but dumped its water into the depths much sooner, in latitudes where the water was warmer and lower in oxygen. But that usually happened in a world that was already a good deal hotter than ours, where there was little or no ice at either pole, and that may be the critical difference: that the MOC behaves differently in a hotter climate, like the one we are heading for.

The evidence is still unclear on whether we run a substantial chance of triggering a Canfield ocean and a greenhouse extinction if we let global warming get out of hand. As with many aspects of this issue, we would only find out for sure when it was too late to do anything about it. But it's the only outcome of the current climate crisis that might convert a massive dieback of the human population into an actual extinction.

CHAPTER SEVEN
Childhood's End

The larger the proportion of the Earth's biomass occupied by mankind and the animals and crops required to nourish us, the more involved we become in the transfer of solar and other energy throughout the entire system. As the transfer of power to our species proceeds, our responsibility for maintaining planetary homeostasis increases, whether we are conscious of the fact or not. Each time we significantly alter part of some natural process of regulation or introduce some new source of energy or information, we are increasing the probability that one of these changes will weaken the stability of the entire system . . . We shall have to tread carefully to avoid the cybernetic disasters of runaway positive feedback or of sustained oscillation . . .

This could happen if . . . man had encroached upon Gaia's functional powers to such an extent that he disabled her. He would then wake up one day to find that he had the permanent lifelong job of planetary maintenance engineer. Gaia would have retreated into the muds, and the ceaseless intricate task of keeping all the global cycles in balance would be ours. Then at last we should be riding that strange contraption, 'the spaceship Earth,' and whatever tamed and domesticated biosphere remained would indeed be our 'life support system' . . . Assuming the present per capita use of energy, we can guess that at less than 10,000 million [people] we should still be in a

Gaian world. But somewhere beyond this figure, especially if the consumption of energy increases, lies the final choice of permanent enslavement on the prison hulk of the spaceship Earth, or gigadeath to enable the survivors to restore a Gaian world.

—James Lovelock,
Gaia: A New Look at Life on Earth, 1979

THIRTY YEARS AGO, when independent scientist James Lovelock wrote his seminal first book about what is now called "Earth System Science" in academic circles, but which he boldly named "Gaia," his concluding remarks about the fate of the "planetary maintenance engineer" struck me so forcibly that I have been able to quote them verbatim ever since. The world in which our species built its civilization used to seem such a stable, welcoming place, and maintained its stability so effortlessly and even invisibly, that nobody in the past would have dreamed of wishing to take up that thankless role, even if they could have imagined that human beings might one day acquire the knowledge and the power to take over the management of the Earth system. But we are now well on the way to acquiring those abilities, at least in a rudimentary form, and it begins to look probable that we will need some of them.

Birth rates have dropped sharply in the three decades since Lovelock wrote his first book. As a result, we are still well short of the ten billion people he set as the point at which our numbers might simply overwhelm the natural systems that regulate the global temperature, the chemical composition of the atmosphere, and other key elements in the equation and thereby succeed (most of the time) in keeping conditions on the planet suitable for abundant life. We may never hit ten billion, and the further short of that destination that we fall, the better it will be. But our energy consumption per capita has increased vastly beyond what anybody in the 1970s could have imagined—

nobody then foresaw the rapidly industrializing Asia of today—and the net effect is about the same. We are overwhelming the natural systems, and rapidly approaching the "runaway positive feedbacks" that concerned Lovelock even so long ago.

In 2007, the Intergovernmental Panel on Climate Change stated that global emissions of greenhouse gases must peak by 2015 if we are to have any chance of keeping the temperature rise to two degrees Celsius (and thus have a reasonable chance of not tripping the feedback mechanisms that could pitch us into runaway heating). In the same year, the International Energy Agency predicted that world energy use will grow 50 percent by 2030, and that fossil fuels will account for 77 percent of that increase. Only instant massive mobilization and wartime-style controls in every major industrialized and industrializing country could stop the rise in greenhouse-gas emissions by 2015, and you know that is not going to happen. So we are going to bust the boundaries. Indeed, the question that looms over us is the same one that comes from the back seat of the family car every ten minutes on long drives: "Are we there yet?" "There" being, in this case, the point at which we have to accept the job of planetary maintenance engineer, at least temporarily—and I think the answer is "Yes."

It was probably already too late to avoid inheriting the job even thirty years ago, although that was not clear at the time because we did not comprehend the sheer momentum of the industrial systems that we have built. It is almost certainly too late now. And maybe it is not altogether a bad thing that the most sentient form of life on the planet is beginning to acquire some ability to regulate the working of Gaia, in our own interests first of all, but potentially in the interests of the entire system.

This statement will stoke instant rage in those to whom rage comes easily, and cause dismay in many more who are appalled by what human civilization has done to the planet in less than ten thousand years. How dare anybody propose human beings as the stewards of the biosphere when their whole history shows that they are just a blight on the planet, devastating the land and emptying the sea of life? What the human race must do is leave everything that remains of the natural world alone to heal as best it can, and tread as lightly as we can upon the Earth. You can't trust us to intervene. Look at our track record.

Again, it's too late. That would have still been a viable course in 1800, but it isn't now. There are six and a half billion of us, and it is almost impossible to imagine a way that we can stop the growth before there are eight and a half billion. Our per capita energy use is immense, and it will continue to grow for at least two generations. The only way that our numbers could come down to a more "sustainable" total in less than several centuries is mass death through famine, war and disease. That may well happen, but I do not want it to: a great deal would be lost, and not just lives. Indeed, if the "wipeout" scenario has any relevance to our situation at all, then you definitely do not want high-technology human civilization to break down, because it provides us with the only set of tools that might enable us to avoid the very worst outcome of our current activities: a full-blown mass extinction.

Yes, of course: technically speaking, we have already initiated an extinction event simply by taking over so much of the planet's surface for our own purposes. The rate at which species are now disappearing is probably at least ten times higher than "normal," and maybe a hundred times higher. (More precise statements of the extinction rate are to be viewed with some suspicion, since nobody knows how many species there are.) But at the risk of sounding unsympathetic, I must point out that species come and go, and that 99 percent of those that ever lived were already extinct before human beings even evolved. For aesthetic reasons, we should stop decimating what is left of the large animal species that remain on the land and in the sea, but our most important priority is to preserve the species that perform vital functions in maintaining the biosphere (most of which are tiny and not in the least cuddly). That is a tricky business, in part because we don't always know which ones they are, but it is much more important than saving polar bears.

To the extent that the biosphere has operated without human intervention to maintain the recent climatic equilibrium—that is, the ten-thousand-year spell of warm and stable climate during which we have built our civilization—we should of course leave it alone to get on with the job. But we should be honest with ourselves: we are actually seeking to preserve one particular climatic state among many potential ones, ranging from deep glaciation to greenhouse extinction, because it suits our particular needs and tastes. We should also be realistic about what needs to be done. We have destabilized this highly desirable climatic equilibrium by our own inadvertent interventions in the past (two hundred years of burning fossil fuels), but given where we are now, it is highly unlikely that we can achieve the goal of restoring that equilibrium without further large and deliberate interventions in the system. We don't know enough yet about how the system works to do that safely, but that doesn't mean that we must never do it. It means that we must learn a lot more about the climate system, very fast, so that we *can* do it more safely.

> By failing to see that the Earth regulates its climate and composition, we have blundered into trying to do it ourselves, acting as if we were in charge. By doing this, we condemn ourselves to the worst form of slavery. If we choose to be the stewards of the Earth, then we are responsible for keeping the atmosphere, the ocean and the land surface right for life. A task we would soon find impossible . . .
>
> To understand how impossible it is, think about how you would regulate your own temperature or the composition of your blood. Those with failing kidneys know the never-ending daily difficulty of adjusting water, salt and protein intake. The technological fix of dialysis helps, but is no replacement for living, healthy kidneys.
>
> —James Lovelock, *The Independent*, January 16, 2006

I see Jim Lovelock as the most important figure in both the life sciences and the climate sciences for the past half-century. I suspect that in another hundred years, if enough from the present survives, he will be granted equal billing with Charles Darwin in the pantheon of scientific heroes. And here, in this quote from *The Independent,* he is saying that we must not do what I am recommending. To which I reply: if I have kidney disease, I definitely want dialysis. It might keep me alive long enough for a more permanent cure to be discovered, and at the least, it will give me more years with those I love.

Getting through the rest of this century without falling into the runaway global warming that Lovelock predicts is only likely if we do not breach the plus-2-degrees-Celsius boundary, and most climate scientists I have spoken to would feel a great deal easier if we never exceeded plus 1.5 degrees. Getting the concentration of carbon dioxide in the atmosphere back down to a relatively safe level—say, the 350 parts per million that James Hansen now advocates as a provisional target—would take even longer, for we passed that milestone some time in the 1980s. We can fully decarbonize our economies if we have enough time, but we are not going to achieve it on the schedule that is required if we are not to breach the boundary. We will pass 400 parts per million of carbon dioxide before 2012, we will probably hit 450 parts per million in the late 2020s, and it will be a miracle if we don't reach 500 parts per million before we can turn the tide.

We (and the biosphere in its current configuration) will only come through this crisis without huge losses if we can keep the temperature from going too high, despite what is happening in the short term to the carbon dioxide concentration in the atmosphere. We are going to get the miserable job of planetary maintenance engineer for a while, but the goal must be to work ourselves out of a job: to restore the natural systems that have done an excellent job of keeping the planet suitable for abundant life most of the time for the past several billion years. This is an emergency, however, and we have to intervene or all of Lovelock's predictions will come true.

What is being proposed is not intervention on a broad front. Nobody is suggesting that we take over the task of providing directly the ecosystem services that are now provided for free by ocean plankton, for example; rather, we should keep the sea surface temperature low enough for them to flourish. We don't know enough about the Earth system to do any fine tuning—but we probably do know how to keep the temperature below two degrees Celsius for the fifty or seventy-five years when we overrun the 450-parts-per-million boundary. If we don't, then we really are screwed.

But just keeping the temperature down artificially will not stop the oceans from becoming more and more acidic due to increased carbon dioxide, I hear a protester cry. No, it won't, but do you have some credible alternative that will stop acidification? Don't tell me "early and steep cuts in emissions," because I don't believe in the Climate Fairy. But we don't have to settle this debate right now. Let's wait five or ten years, and if those "early and steep cuts in emissions" still haven't happened, then we'll discuss it again. There's enough time for that.

In the meantime, though, I'd like lots of research to be done on geo-engineering techniques of all kinds, because I suspect that we will need them. A high level of carbon dioxide in the atmosphere has undesirable consequences that just keeping the temperature down cannot stop, but it would help a lot if we could keep the average global temperature low enough to avoid triggering large natural feedbacks that take the situation completely beyond our control. Keeping the temperature down could also prevent the kind of human catastrophes, including great wars, that would doom all our efforts to clear up the mess we have made.

The job, for the rest of this century, is repairing the damage we did over the past two centuries of industrialization to the homeostatic, Gaian systems that we didn't even realize we depended upon until relatively recently. That does not mean that we de-industrialize—this global society will live or die as a high-energy enterprise—but to begin with we must completely de-carbonize our energy, our transportation, and our industry. Then much of the forests that we cut down over the past two hundred years must be replanted, huge no-fishing reserves must be created to permit the repopulation of the oceans, and the amount of land we have removed from the natural cycles in order to grow food on it must fall from the current 40 percent of the Earth's land surface to 30 percent or less. It's not too late to fix most of the damage, if we have enough time and are not fatally distracted by catastrophes.

And how do we feed the eight-and-a-half billion people of the mid-century world—or nine billion, or nine-and-a half, take your pick—while we are reducing the amount of land under cultivation and cutting back on fishing? This is a high-tech civilization, and I suspect that the answer lies in that direction. Alberta, for example, produces more wheat and beef than any other Canadian province, but it is approaching the limit in terms of available water. So it is looking into non-traditional ways of producing those commodities.

> We really should be getting beyond growing the whole plant or the whole cow, when only certain parts are of primary value to us. If you think about wheat, it is primarily the kernel of wheat that is of primary importance to us . . . There are significant advances in molecular biology and in cell biology. In the medical field, we're really quite far advanced in growing artificial skin. You start with a small patch of human skin, and you grow a large patch. There's no real reason to think that that cannot be extended to other parts of the human body . . . and if that is so, it isn't all that far-fetched to think, well, why don't we just find ways of

growing steak, or fish fillets, from a few cells by providing the appropriate nutrient medium?

What triggered this exploration in part was our discussion about water. In order to grow one kilogramme of beef, you require about thirteen thousand litres, thirteen thousand kilogrammes of water. The same is true for plants. Conceivably you can grow just the kernel, or just the starch that makes up the kernel, artificially . . . To grow a kilogramme of wheat flour takes about a thousand litres of water. If you want to visualize this as a milk carton of one litre, it takes about three bathtubs of water to produce that one milk carton of wheat flour. That's a very profligate way of using water, and with impending climate change, concerns about droughts, it's reasonable to try to explore other ways, not necessarily as replacements but as complements. And I think modern science is putting it within reach now . . .

It is not unnatural for a modern variant of an industry to have its roots in a traditional variant, so I could certainly foresee in the years to come that there will be farmers who could produce wheat or beef in both ways simultaneously, possibly for different markets.

—Axel Meisen, chair of foresight, Alberta Research
Council, in an interview with the author, May 2, 2008

Many people will simply be horrified by this proposal, and many others will suspect that the innocent little phrase, "possibly for different markets," foreshadows a two-tier world food system: real beef for the rich, vat-grown beef for the poor. Even if their taste and texture were exactly the same, considerations of prestige would produce that two-tier market. Still, it's a lot better than *Soylent Green*, and it could offer a way to uncouple human food production from the traditional farming techniques that have led us to alienate 40 percent of the world's surface—and the most productive 40 percent, at that—from its proper task of contributing to the maintenance of the biosphere. We are going to see more proposals like this, and we are not going to be able to dismiss them out of hand.

Let us make the heroic assumption, just for a moment, that the human race will be clever enough to make it through this century without triggering runaway warming and a massive population dieback. Let us further assume that we have retained our high-technology civilization (for otherwise our chances of making it through unscathed would be very small), and that the experience has taught us something about the need to respect the natural systems that we depend upon. You may see these as low-probability assumptions, but you cannot deny that they are at least possible. What would that somewhat chastened end-of-the-century global society look like?

It would be a world with much greater equality of wealth between the old rich countries and the Majority World, because that is a precondition for making it through the crisis. Even with the most stringent population controls there would probably still be five or six billion of us, although there might be a gradual downward trend. Since most of those five or six billion would have access to the full industrialized lifestyle, enormous emphasis would have to be put on learning to "live lightly on the planet." Given the right technologies, it is not improbable that most people would still have personal transportation devices of some sort, that long-distance travel would continue to be possible for more than the privileged few, that those who wished to would still be eating meat (although, in many cases, ethically produced, vat-grown meat). This is not a wish-fulfilment dream; it is what we would probably get if we pass the test.

Various metaphors for our present situation come to mind, but the one that really sticks is the final exam. For more than ten thousand years, human beings have built a civilization that is now global in extent, but for most of that time we were really semi-barbaric children. Only two centuries ago, slavery was almost universal, women were an inferior caste almost everywhere, and war was the normal way of doing business. Resources were always scarce, so competition was usually a better strategy than cooperation. And, until the very end of that period, we had no real comprehension of the workings of the planet we lived on.

Then we began burning fossil fuels, and resources became abundant. Population and consumption both soared, of course, but so did science and knowledge. We began to understand our place in the universe, and that was very frightening. The nursery world that we thought we lived in, half playground, half battle-field, but unchanging and specially designed for human beings, turned out to be a fantasy. The real world was immensely old, it cared nothing for us, and there were many ways it could hurt us that we had never even imagined: ten-thousand-year volcanic cataclysms and hundred-metre sea level changes, ice ages and asteroid strikes, runaway greenhouse warming and supernovas a hundred light years away that could sterilise the planet. We realized we were on our own, and it was time to grow up.

We haven't done all that badly, really. We began by trying to behave better towards one another in our own societies—the great democratic revolutions, free universal education, the invention of the welfare state—and by the end of this two-century period we had even created semi-functional international institutions. Nobody would have put it this way at the time, but with hindsight you can see that we were actually building our capacity to take responsibility for people and events beyond our own horizons. Just as well, given what lies in wait for us. But it is worth remembering that all this only became possible because large numbers of people finally had the security and the leisure to think beyond the moment and to act for the future. We owe a lot to fossil fuels.

There was no alternative to burning fossil fuels in terms of getting an industrial, scientific civilization off the ground, because no other source of energy was available to a low-technology society. (And it was a one-time-only offer: we have used up all the easily accessible sources of fossil fuel, and any descendants of ours who are trying to restart an industrial civilization will be out of luck.) We went on burning coal and oil and gas heedlessly for almost two centuries, not suspecting that, in the long run, dependence on fossil fuels is a kind of suicide pact. And here is the little miracle that shows we still have more than our share of luck: at exactly the same time when it became clear that we have to stop burning fossil fuels, a wide variety of other technologies for generating energy became available. We are truly blessed.

So now we have to manage the transition, and we have about half a century to complete the job. Most of the changeover has got to come in the next twenty years, and we need to have completely decarbonized our economies by 2050. In the meantime, we have to keep the global average temperature from passing the plus-two-degrees-Celsius boundary no matter what the carbon dioxide concentration in the atmosphere is doing, and in the longer run we need to get the carbon dioxide down to 350 parts per million. That won't be easy, but it is the sort of task at which industrial societies excel.

We just barely scraped through the mid-term exam in the last century: we acquired the ability to destroy our civilization directly, by war, and we managed not to use it. Now it's the final exam, with the whole environment that our civilization depends on at stake. It's not just about knowledge and technical ability; it is also about self-restraint and the ability to cooperate. Grown-up values, if you like. How fortunate that we should be set such a test at a point in our history where we have at least some chance of passing it. And how interesting the long future that stretches out beyond it will be, if we do pass.

Acknowledgements

MY THANKS, FIRST OF ALL, to the dozens of busy people who took time to give me the interviews that have done so much to shape this book. All opinions, interpretations, and errors are, of course, my own.

Melissa Howells did much of the research, including the job of finding and setting up interviews with the right people. As so often, I am in her debt for much more than I paid her. And my love and thanks to Tina Viljoen, who gave indispensable advice on everything from content to punctuation at every stage of the process.

Permissions

Grateful acknowledgement is made to the following for the permission to reprint previously published material.

p. 5 "The future effects of climate change will stem from a more unstable process, involving sudden and possibly in some cases catastrophic changes . . ."
From *DCDC Global Strategic Trends Programme 2007–2036*, Third Edition, pp. 26–28. Ministry of Defence, London. Reprinted by permission of the Ministry of Defence, DCDC, Shrivenham, SWINDON, United Kingdom.

p. 9 "You already have great tension over water [in the Middle East] . . ."
From *National Security and the Threat of Climate Change* The CNA Corporation, Noel L. Gerson Vice President, Communications and Public Affairs. Reprinted by permission of The CNA Corporation.

p. 10 "People are saying they want to be convinced, perfectly . . ."
From *National Security and the Threat of Climate Change* The CNA Corporation, Noel L. Gerson Vice President, Communications and Public Affairs. Reprinted by permission of The CNA Corporation.

pp. 15–16 "[This scenario] assumes that the [IPCC 2007 report's] projections of both warming and attendant impacts are systematically biased low."

From "The Age of Consequences: The Foreign Policy and National Security Implications of Global Climate Change," Center for Strategic and International Studies, 2007, Project Co-Directors Kurt M. Campbell, Alexander T J Lennon, Julianne Smith. Reprinted by permission of the Center for Strategic and International Studies and the Center for a New American Security.

pp. 27–28 "The Earth has recovered from fevers like this (in the past) . . ."

From *The Revenge of Gaia*, James Lovelock, published by Allen Lane, the Penguin Group, London, 2006, p. 60. Reprinted by permission of James Lovelock and Allen Lane, the Penguin Group.

p. 43 "One of the things that struck me . . ."

From *National Security and the Threat of Climate Change* The CNA Corporation, Noel L. Gerson Vice President, Communications and Public Affairs. Reprinted by permission of The CNA Corporation.

p. 61 "Northern Eurasian stability could . . . be substantially affected by China's need to resettle many tens, even hundreds of millions from its flooding southern coasts . . ."

From "The Age of Consequences: The Foreign Policy and National Security Implications of Global Climate Change," Center for Strategic and International Studies, 2007, Project Co-Directors Kurt M. Campbell, Alexander T J Lennon, Julianne Smith. Reprinted by permission of the Center for Strategic and International Studies and the Center for a New American Security.

p. 67 "Paleoclimate data shows that climate sensitivity is approximately 3 degrees Celsius for doubled CO_2, including only fast feedback processes . . ."

From the abstract of "Target Atmospheric CO_2: Where Should Humanity Aim?" by James Hansen, Makiko Sato, Pushker Kharecha, David Beerling, Valerie Masson-Delmotte, Mark Pagani, Maureen Raymo, Dana L. Royer and James C. Zachos; http://arxiv.org/abs/0804.1126, Submitted by James Hansen. Reprinted by permission of James Hansen.

pp. 103–104 "Abu Dhabi is developing nearly 30,000 hectares of farmland in Sudan in the first step towards ensuring food security in the emirate . . ."

From "Abu Dhabi develops food farms in Sudan" by Xan Rice, in *The Guardian*, Wednesday July 2, 2008. Reprinted by permission of *The Guardian*.

pp. 108–109 "The Chinese navy is poised to push out into the Pacific—and when it does, it will quickly encounter a US Navy and Air Force unwilling to budge from the coastal shelf of the Asian mainland . . ."

Robert D. Kaplan, "How We Would Fight China: The Next Cold War," *Atlantic Monthly*, May 2005.

p. 113 "In December 2001, when terrorists attacked the Indian parliament, India blamed Pakistan and withdrew her High Commissioner in protest . . ."

From Part II, Chapter 6 of "The Final Settlement: Restructuring India-Pakistan Relations," Strategic Foresight Group, International Centre for Peace Initiatives, Mumbai, 2005.

pp. 134–135 "Because our televisions and our lights were burning already, the system was straining even before these extra demands were made . . ."

From *Heat: How to Stop the Planet Burning*, George Monbiot, London, 2007, Penguin Books, pp 80–81. Reprinted by permission of George Monbiot and Penguin Books.

p. 187 "The people in the Department of Energy have done everything possible to block decentralised power. . . ."
   Ken Livingstone, Mayor of London, 2000–08. Interview by Jo Revill, Whitehall editor in *The Observer*, Sunday March 23, 2008. Reprinted by permission of *The Guardian*.

p. 189 "Many members of the scientific and technical communities fear that the full effects of various geo-engineering schemes are not fully understood . . ."
   From Wikipedia, "Geo-engineering," entry as of June 3, 2008 http://en.wikipedia.org/wiki/Planetary_engineering.

p. 195 "Recent research has shown that the warming of the Earth by the increasing concentration of CO_2 and other greenhouse gases is partially countered by . . . sulfate particles, which act as cloud condensation nuclei . . ."
   "Albedo Enhancement by Stratospheric Sulfur Injections: A Contribution to Resolve a Policy Dilemma?" by Paul Crutzen in *Climatic Change* (2006) 77: 211–12, published by Springer. Reprinted with permission from Paul Crutzen and Springer.

pp. 225–226 "Simple calculations show that if deep-water H_2S concentrations increased beyond a critical threshold during oceanic anoxic intervals of Earth history . . ."
   From the abstract of "Massive release of hydrogen sulfide to the surface ocean and atmosphere during intervals of oceanic anoxia," by Lee R. Kump, Alexander Pavlov, and Michael A. Arthur, *Geology*; May 2005; v. 33;no. 5; p. 397–400 http://geology.geoscienceworld.org/cgi/content/abstract/33/5/397.

pp. 228–229 "No wind in the 120-degree (F) morning heat, and no trees for shade. There is some vegetation but it is low, parched, stunted . . ."

From *Under a Green Sky*, Peter D. Ward, New York 2007, Smithsonian Books, HarperCollins Publishers, pp.139–40. Reprinted with permission from Peter D. Ward and HarperCollins.

p. 233 "The larger the proportion of the Earth's biomass occupied by mankind and the animals and crops required to nourish us, the more involved we become in the transfer of solar and other energy throughout the entire system . . ."

From *Gaia: A New Look at Life on Earth*, James Lovelock, Oxford, 1995, Oxford University Press, pp. 123–24. Reprinted with permission from James Lovelock and Oxford University Press.

pp. 237–238 "We have given Gaia a fever and soon her condition will worsen to a state like a coma . . ."

From "James Lovelock: The Earth is about to catch a morbid fever that may last as long as 100,000 years," *The Independent*, January 16, 2006. Reprinted with permission from *The Independent* and James Lovelock.

Every effort has been made to contact the copyright holders; in the event of an inadvertent omission or error, please notify the publisher.

Index

GWYNNE DYER has served in the Canadian, American and British navies. He holds a Ph.D. in war studies from the University of London, and has taught at the Canadian Forces College and the Royal Military Academy Sandhurst. Dyer is the author of the bestselling and now classic *War*, and the writer and host of the Oscar-nominated NFB-CBC television series of the same title. He currently writes a syndicated column that appears in more than 175 newspapers around the world; the best of his most recent writing has been collected in *With Every Mistake*.